Phased Array Antenna Analysis

Using Finite Hybrid Element Methods

BY

Daniel T. McGraith

Wexford Press
2008

Table of Contents

List of Figures

List of Tables

List of Symbols

η_0	Impedance of free space (real, constant) [Ohms]
θ	Spherical coordinate angle measured from the $+z$ axis [radians]
θ_0	Array scan angle (real constant) [radians]
λ	Wavelength (real constant) [m]
μ	Permeability (material physical property, complex scalar) [Henrys/m]
μ_0	Permeability of free space (real, constant) [Henrys/m]
$\bar{\xi}_s$	Two-dimensional Fourier transform of component of expansion function for edge s tangential to radiation boundary
ϕ	Spherical coordinate angle measured from $+x$ axis [radians]
ϕ_0	Array scan angle (real constant) [radians]
ψ_x, ψ_y	Beam steering phase shifts (real constants) [radians]
$\bar{\psi}_s$	Vector expansion function associated with mesh edge s
Ψ_{si}	Inner product of expansion function $\bar{\psi}_s$ and waveguide mode function $\bar{g}_i$
Ω	Volume region, usually the unit cell excluding interiors of conductors
ω	Radian frequency (real constant) [radians/s]

Abstract

This research in computational electromagnetics addressed the problem of predicting the near-field mutual coupling effects in phased array antennas. It developed and demonstrated a new analysis technique that uses the finite element method (FEM) in combination with integral equations. Due to FEM's inherent ability to model inhomogeneous dielectrics, the new capability encompasses many radiator types that were not amenable to analysis by previously-existing methods. The analysis considers the general case of a radiator in an infinite array that is fed through a ground plane by one of three types of waveguides: rectangular; circular; or circular coaxial. Accurate feed modeling is accomplished by enforcing continuity, between the FEM solution and an arbitrary number of waveguide modes, across the ground plane aperture. A periodic integral equation is imposed at a plane above the antenna's physical structure to enforce the radiation condition and to confine the analysis to a single array unit cell. The electric field is expanded in terms of vector finite elements, and Galerkin's method is used to write the problem as a matrix equation. The Floquet condition is imposed as a transformation of the matrix, which is equivalent to wrapping opposing unit cell side walls onto each other with a phase shift appropriate to the scan angle and lattice spacings. The solution of the linear system, accomplished using the conjugate gradient method, gives the electric field, from which the active reflection coefficient and active element gain are calculated.

The theory and formulation were used to develop a general-purpose computer code. The use of commercial CAD (computer-aided design) software for geometry and mesh generation makes the code geometry independent. It was validated by comparing its results to published data for arrays of open-ended waveguides, monopoles and microstrip patches. Predictions for dielectric-clad monopoles were validated by a hardware experiment. Finally, the code was used to predict the scanning properties of arrays of printed dipoles and printed flared notches.

PHASED ARRAY ANTENNA ANALYSIS

USING HYBRID FINITE ELEMENT METHODS

I. Introduction and Background

1.1. Introduction

A critical problem in phased array antenna design is that of controlling the *mutual coupling* between individual antennas, or *radiators*, that comprise the array. Mutual coupling may reduce antenna efficiency by creating a reflection mechanism that depends on both the radiator geometry and the scan angle. The limiting case, but one that frequently results from poor radiator design is *scan blindness*, meaning that there are one or more angles in the desired field of view where the reflection is total. Since mutual coupling is inherently a near-field electromagnetic effect, it is rarely possible to achieve an acceptable radiator design without the ability to accurately model the near fields. As we attempt to design phased array antennas for new applications, we often find that existing field computation techniques, most notably moment methods, do not adequately account for the radiator's topology, feed structure, or dielectric materials. This is especially true for several broadband radiators that incorporate dielectrics either as structural support or as electrical loading that reduces their size. Therefore, improved computation techniques are required for use as design tools for broadband phased array antennas.

The objective of this dissertation research was to develop and demonstrate a new analysis method versatile enough to predict the performance of a variety of radiators. The method is a hybrid of the finite element method (FEM) with integral equation continuity conditions. The integral equation for one of three types of waveguides (a summation over dominant and higher-

1

order waveguide modes) provides an accurate method for including feed structure effects. The use of a periodic integral equation to represent the fields above the array enforces the radiation condition and makes the analysis tractable by confining it to a single array unit cell. A new periodic boundary condition for finite elements was derived to account for the mutual coupling across unit cell side walls. The work culminated in a general-purpose computer code that successfully predicted the scan-dependent properties of a variety of arrays such as open-ended waveguides, microstrip patches and printed flared notches. When possible, the results were confirmed by comparisons to data in the scientific literature that was obtained from either measurements or by other methods of calculation. These validation cases are only a sampling of the capability of this new analysis tool, which is able to predict the scanning performance of most array radiators that are in use or proposed for use. Its future use as a design tool is bound to improve the performance of phased array antennas.

1.2. Phased Array Antenna Electromagnetic Analysis

The most important reason for near-field electromagnetic analysis of phased arrays is impedance matching, which translates directly into radiation efficiency and low VSWR (voltage standing wave ratio) to prevent damage to transmitter components in high-power applications. Paradoxically, radiating elements that are well matched in isolation do not necessarily remain so when they are arranged in a closely-spaced lattice to form an array. Taking one radiator to be a reference element, some of its transmit power couples directly into other elements, and some power from each of the other elements will likewise couple directly into the reference element. The power returning to the transmitter by way of mutual coupling represents a reflection mechanism. In order to achieve a good *active impedance match*, each element must, by itself, be slightly mismatched, so that its self-reflection vectorially cancels the sum of couplings from other elements. There are numerous methods for controlling these mismatches, such as altering the

2

radiator geometry or including matching circuits in the feed network, but they can only be effective if an accurate solution for the mutual coupling is available [1:1662].

Most existing solutions for array mutual coupling use the method of moments (MoM) in conjunction with an infinite array approximation. The approximation makes the problem computationally tractable since periodicity conditions may be used to restrict the analysis to the space around a single radiator, called a *unit cell*. It is a reasonable approximation for large arrays, and it is also used as part of the present method.

1.3. The Need for Improved Analysis Methods

There are several trends in antenna development that are causing array element designs to outstrip the available analysis methods. One is the desire to use electronic scanning antennas in applications such as airborne satellite communications and airborne surveillance radar, in which the antennas' intrusion, protrusion and weight must be minimized [2]. Another is the growing trend toward millimeter wave frequencies, leading to dimensions so small that it is impractical to fabricate and assemble and array one element at a time. Yet another is the potential cost reduction of using printed circuit and integrated circuit, or *monolithic*, fabrication [3].

The radiating elements that are proposed to meet these new requirements often include irregularly-shaped conducting surfaces in combination with inhomogeneous dielectrics. An example, shown in Figure 1a, is the "flared-notch" element [4]. The dielectric card is clad with metal on the back side except for a slot that opens progressively wider near the top. The front side is bare except for a microstrip feed line. It is shown here fed from a coaxial transmission line that penetrates a ground plane, but the feed line could be microstrip as well. A recent MoM analysis of this radiating element in the phased array environment used the simplification shown in Figure 1b: the dielectric is ignored; and the feed line is replaced by an delta-gap source across the slot [5]. It is clear that this model cannot predict the effects of high-dielectric-constant sub-

3

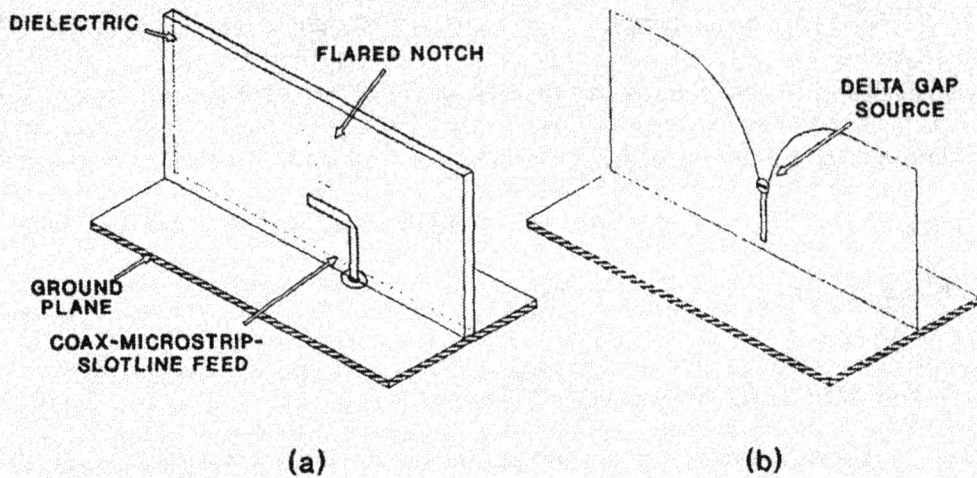

Figure 1. Flared Notch Radiator: (a) Printed Circuit Fed from Coaxial Waveguide;
(b) Geometry Model for Method of Moments

strates (monolithic antennas commonly use Gallium Arsenide, whose relative permittivity is 12.8-

12.9) or of radiation from the feed structure.

A second example, shown in Figure 2a is a dipole radiator metallized on one side of a

dielectric card, with a balun feed metallized on the other [6]. The dielectric may be trimmed or

notched at each end of the dipole to reduce the mutual coupling between elements located at

intervals along the card. Figure 2b is the structure actually modeled using an innovative MoM

approach. In this case, the presence of the dielectric was taken into account by using a Green's

function for a parallel-plate region periodically loaded with dielectric slabs [7],[8]. Some results

of that work confirm that the dielectric has a pronounced effect on the wide angle scanning

properties. On the other hand, the sweep-back of the dipole arms, which is known to be impor-

tant for achieving a good impedance match at wide scan angles [9], is neglected. Also, the feed

is modeled as a simple delta-gap source, neglecting the effects of the balun.

4

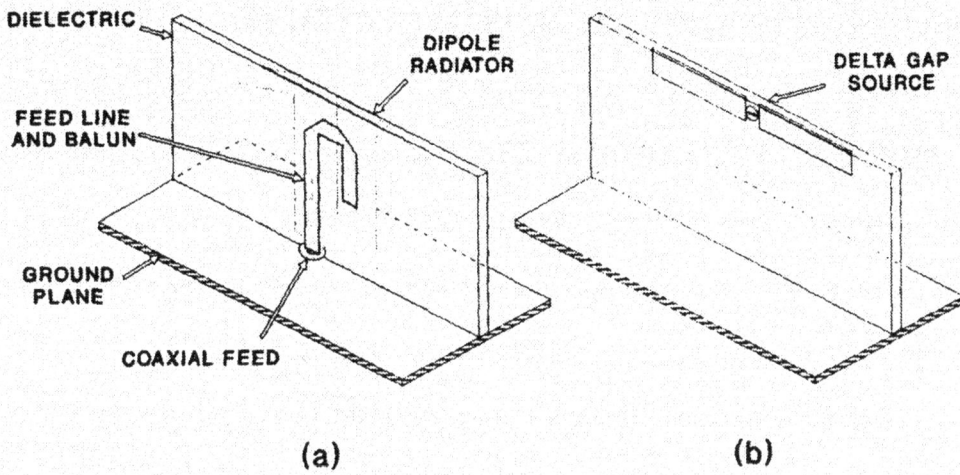

Figure 2. Printed Dipole Radiator: (a) Actual Geometry with Microstrip
Balun and Coaxial Feed; (b) Method of Moments Model

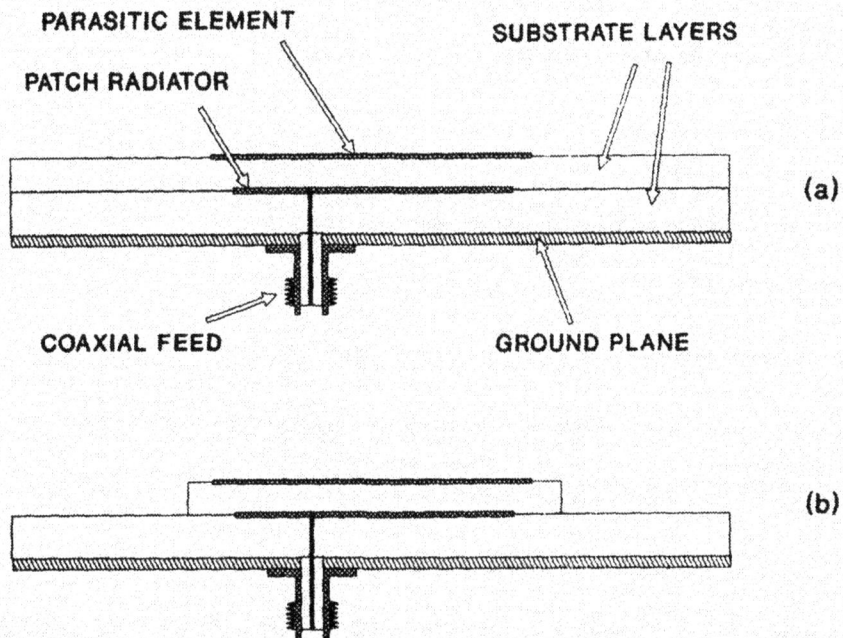

Figure 3. Stacked Patch Radiator with Coaxial Feed: (a) Two Continuous
Substrate Layers; (b) Non-continuous Top Substrate

5

A third and final example is shown in Figure 3a. It is essentially a rectangular microstrip patch, fed through the ground plane from a coaxial cable. The second "proximity-coupled" patch on top of the upper dielectric layer is intended to increase the overall bandwidth beyond the 5% [10] that is typical of a single patch. The array properties of this radiating element have been predicted using MoM [11], with the coaxial feed represented by a frill current source. The moment method approach for this, and other microstrip problems, relies on using Green's functions for layered, infinite dielectric slabs. Those Green's functions do not apply to situations such as Figure 3b, in which one or both substrates are not continuous layers.

These three examples illustrate the deficiencies in previous mutual coupling computation techniques; and the corresponding new capabilities that have been obtained with the hybrid finite element method: (1) multiple dielectrics with arbitrary shape; (2) irregular conducting surfaces that support currents flowing in arbitrary directions; and (3) detailed feed structures. A further objective that is also important is geometry independence: Whereas each of the three examples discussed above used a specialization of MoM to the particular structure and required development of a separate computer code, the present work resulted in a code that can model all three, and many others as well.

1.4. Methods in Computational Electromagnetics

The techniques of "classical electromagnetics" provide formalisms for casting physical problems as mathematical boundary value problems. Since the solutions that may be obtained by the purely analytical methods are restricted to canonical geometries, most electromagnetic design problems of current interest require the use of numerical methods to obtain a solution. The techniques of "computational electromagnetics" (CEM) are formalisms for mapping boundary value problems from continuous to discrete forms so that they may be solved by computer. Therefore, the objective of CEM is to produce tools, i.e. computer codes with which device

6

designs may be evaluated without resorting to hardware experiments. The tools must have four essential characteristics: *effectiveness, reliability, efficiency*, and *versatility*. In other words, they must consistently obtain correct results at reasonable cost for a variety of problem geometries. The previous section showed that the shortcomings of current methods for phased array near-field analysis are mainly in effectiveness (due to simplifications of the actual problem geometry) and versatility (due to the restriction of each to very specialized geometries).

The requirement for effectiveness limits the search for new techniques to integral equation (MoM) and partial differential equation (PDE, finite element and finite difference) methods. The former is by far the most well advanced for solving antenna problems because the integral equations incorporate Green's functions that satisfy radiation conditions, forcing all valid solutions to decay to zero with increasing distance from the field sources (equivalent currents). The PDE techniques, on the other hand, more easily account for dielectric inhomogeneities for which Green's functions are not available. The finite element method is more appropriate for devices with irregular, especially curved surfaces, because it may use irregular grids, or "meshes," while the finite difference method typically uses regular, Cartesian grids. A similar need to model objects that include inhomogeneous dielectrics has led researchers in electromagnetic scattering to consider hybrids of the finite element method with integral equation methods [12]-[14]. By surrounding the object with an imaginary boundary in free space surrounding the scatterer, the finite element method may be used to solve for the fields inside the boundary as though it were an enclosed region, and an integral equation is imposed to ensure field continuity across the boundary. Hence, the hybrid finite element method (HFEM) appeared to be a likely choice for the phased array radiator problem as well, provided that a means could be found for implementing periodic boundary conditions. The success of that implementation and a demonstration of its benefits are some of the important results of this dissertation research.

The next chapter will present an overview of the solution approach, which involves three novel continuity conditions for the three dimensional finite element problem. The detailed analyses and derivations constituting the problem "formulation" (its description as a mathematical boundary value problem, and its reduction to a linear system of equations) are given in Chapters III-VI. They are intended as documentation, or as a trail for the reader who would attempt a similar solution to related electromagnetic problems. They are not essential to understanding the results of validation tests and hardware experiments presented in Chapters VII-X.

This hybrid finite element method is a frequency-domain approach. Hence, throughout this document, all field and current quantities are understood to be time-harmonic, with the complex exponential $e^{j\omega t}$ time dependence suppressed.

II. Solution Overview

2.1. Problem Description

The generic problem geometry is shown (cross-section) in Figure 4. The "interior region," denoted Ω, is a section of an array unit cell. It is bounded by the surface denoted Γ, whose bottom wall is the ground plane plus the feed waveguide aperture. The top wall, called the "radiation boundary," is an imaginary constant-z surface at an arbitrary position in free space above the radiator structure. The side walls conform to the unit cell boundaries. For example, Figure 5 shows two arrays (stacked-patch radiators), one with a rectangular lattice and the other with a triangular or "skewed" lattice. In each case the unit cell is a cylinder extending indefinitely in the $\pm z$ directions. Its side boundaries are chosen to satisfy the periodicity conditions (discussed in Chapter VI and Appendix C). The region Ω is formed by simply truncating the unit cell at some plane above the array.

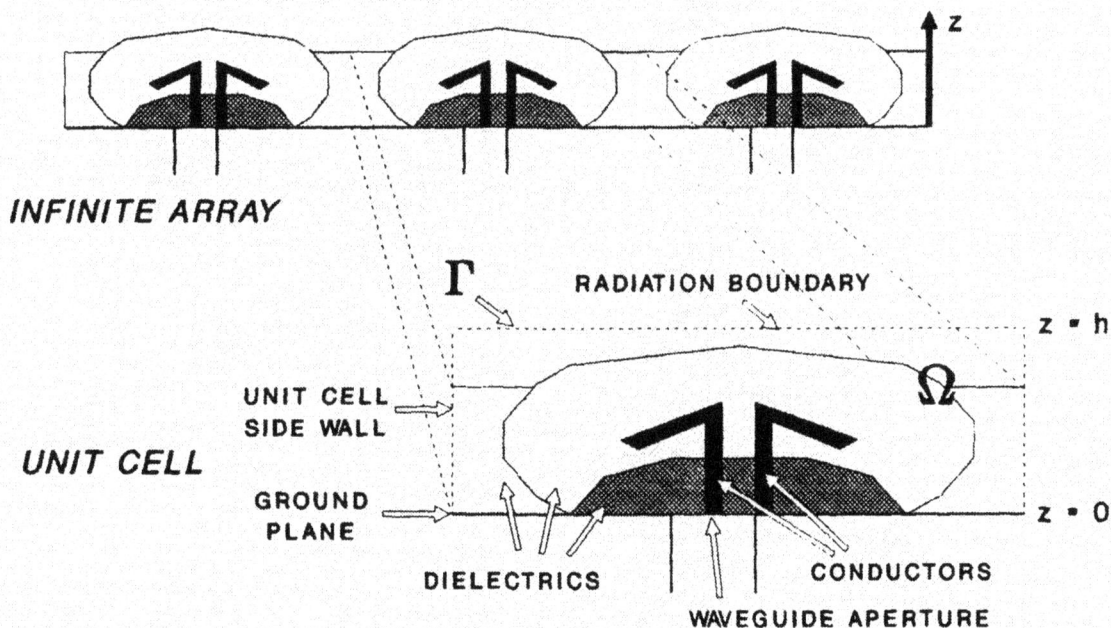

Figure 4. General Phased Array Radiator Problem

9

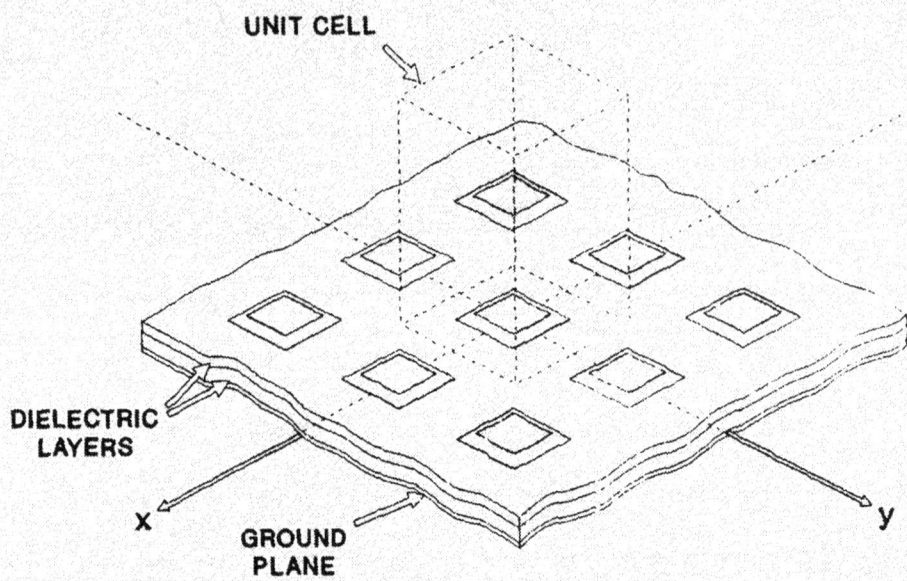

(a)

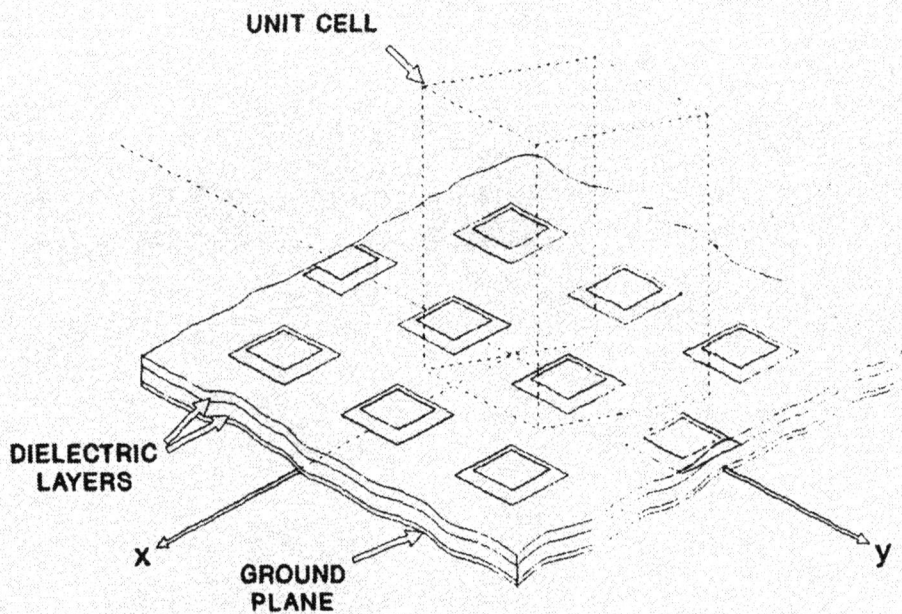

(b)

Figure 5. Unit Cells (Stacked Patch Arrays): (a) Rectangular Lattice;
(b) Skewed (Triangular) Lattice

10

Ω may now be viewed as a cavity containing any number of material regions, each with distinct constitutive parameters ϵ and μ, which may be complex (lossy). This research will only consider linear and isotropic materials. There may also be voids within Ω that represent the interiors of perfectly conducting obstacles. Infinitely thin wires and open surfaces such as patches and strips are also permitted.

2.2. The Matrix Equation

The solution to the boundary value problem represented by Figure 4 is the electric field $\bar{E}(x,y,z)$ everywhere inside Ω and on Γ. HFEM will find an approximation to $\bar{E}$ in terms of piecewise-continuous expansion functions weighted by a column vector of coefficients, denoted **E**. Those coefficients are the solution to the matrix equation

$$\left[S^{EE} + S^{EJ} \right] E = E^{inc} \tag{1}$$

Complete explanations of the terms in this equation are given in succeeding chapters, but briefly: The matrix S^{EE} is sparse, representing local interactions between field sources inside Ω; S^{EJ} represents interactions between field sources on the nonconducting parts of Γ through integral equations. The right side vector E^{inc} is due to a field incident on Γ from the feed waveguide. The performance parameters that are of greatest interest are the active reflection coefficient and active element gain, which may be found directly from those parts of **E** on the waveguide aperture and radiation boundary, respectively.

2.3. Finite Elements

The region Ω will be subdivided into small volume elements. Four-sided tetrahedra were chosen because they conform more readily to irregular and curved surfaces than other popular choices such as six-sided "bricks." The volume elements are often referred to as *cells*, and their four vertices as *nodes*. The collection of tetrahedra is called the *mesh*. Material properties will

11

be assumed constant within each cell. Figure 6 is an example mesh (one quadrant of a thick disk) that illustrates an important flexibility of tetrahedron meshes: the mesh density may be varied within an object. Although 10-20 nodes per linear wavelength is usually an adequate sampling rate, one may wish to sample finer in regions where the field is expected to have singularities; or in order to capture fine details of an object's geometry.

Finite elements are polynomial functions that are defined over individual cells, or *sub-domains*. Chapter III will give a more detailed description of the linear, vector finite elements used in this work. These functions are used as expansion and testing functions for Galerkin's method, which is the mechanism used to reduce the boundary value problem to a matrix problem.

2.4. The Weak Form Functional

The principal distinction between finite element and moment methods (as the terms are commonly used within the electromagnetic research community) is that the former is applied to variational statements, while the latter is applied to integral equations [15:16]. The variational

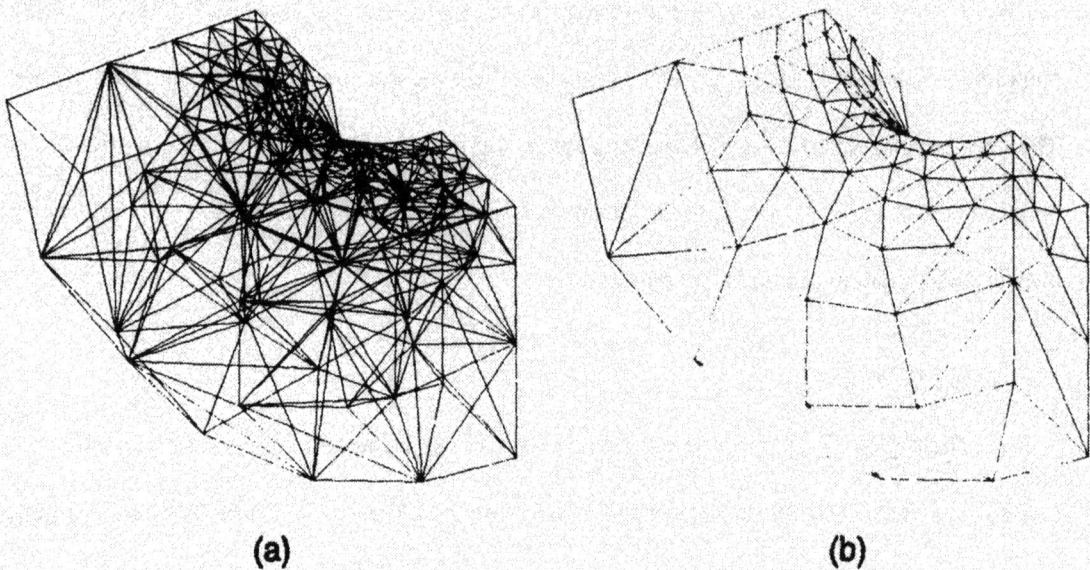

(a)   (b)

Figure 6. Subdivision of a Volume Region into Tetrahedra (one quadrant of a disk):
(a) Interior Edges Visible; (b) Interior Edges Hidden

statement used here is the *weak form* of the vector wave equation. The time-harmonic form of the wave equation for electric fields in a source-free, inhomogeneous region is:

$$\nabla \times \frac{1}{\mu_r} \nabla \times \bar{E} - k_0^2 \epsilon_r \bar{E} = 0 \qquad (2)$$

A functional is constructed by taking its inner product with a trial function, $\bar{W}$, then applying a Green's identity (see Appendix A):

$$F(\bar{E}) = \int_\Omega \left[\frac{1}{\mu_r} \nabla \times \bar{W}^* \cdot \nabla \times \bar{E} - k_0^2 \epsilon_r \bar{W}^* \cdot \bar{E} \right] dv + jk_0 \eta_0 \int_\Gamma \bar{W}^* \cdot \bar{J} ds = 0 \qquad (3)$$

This is called a weak form because the Green's identity has shifted one derivative from the field $\bar{E}$ to the trial function $\bar{W}$, thus weakening the differentiability requirement on $\bar{E}$. This functional has three difficulties that integral equations do not:

(a) the Helmholtz equation specifies the curl only and not the divergence. There-fore, extra effort is required to enforce the divergence condition $\nabla \cdot (\epsilon \bar{E}) = 0$.

(b) Boundary conditions are not included. Thus, although (3) is not restricted to a class of problems with uniform boundary conditions, extra effort is required to ensure the satisfaction of all boundary conditions that are present.

(c) The radiation condition is not enforced.

Chapter III will discuss how (a) and (b) are resolved by choosing vector expansion functions that obey the divergence condition and that satisfy boundary conditions at both perfect conductors and dielectric interfaces. Chapters IV and V show how the radiation condition is enforced by substi-tuting an integral equation for $\bar{J}$ into the boundary integral of (3). In the case of the waveguide aperture, that integral equation will take the form of a sum over waveguide modes. In the case of the radiation boundary, it will be a periodic integral equation, which may be written as a sum over spectral domain sample points, i.e. *Floquet modes*.

13

The last detail to be addressed is the problem of enforcing periodicity conditions at the unit cell walls. The implementation is straightforward in theoretical terms. It has been accomplished by others for two-dimensional grating problems [16]. Chapter VI discusses its extension to three dimensions and the means for implementing it algorithmically: The matrix is first constructed as though the unit cell walls are open circuit boundaries; then the matrix is modified, creating some new terms and removing others. The effect is as if opposing mesh side walls were folded around onto each other with a phase shift appropriate for the array scan angle and the unit cell dimensions.

2.5. Development Approach

Figure 7 illustrates a generic "cavity array problem," a somewhat simpler problem than the "general array" problem of Figure 4. It is still a phased array antenna, but the radiators are separated from each other by conducting walls. Their mutual coupling is only through apertures in a conducting ground plane. This is appropriate to a restricted class of radiators such as open-

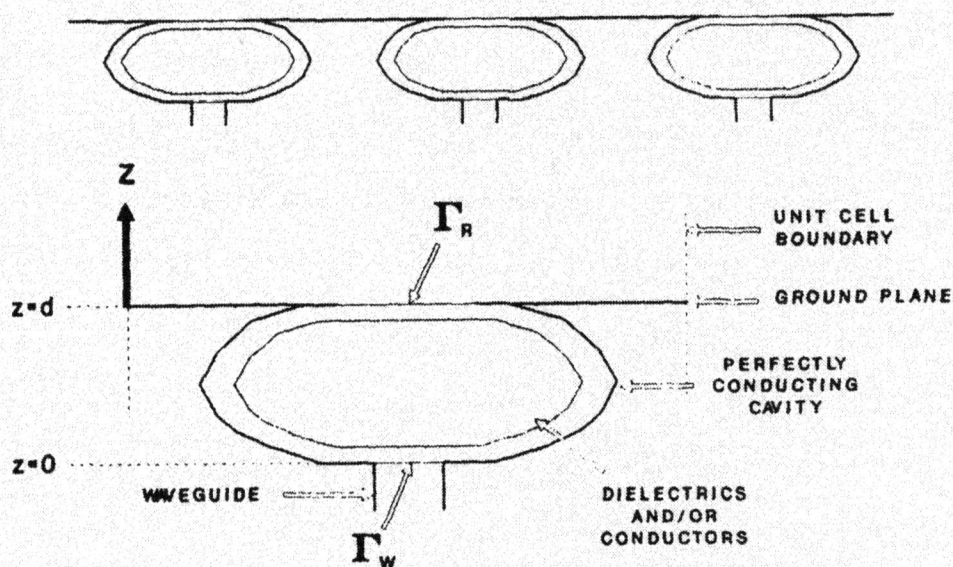

Figure 7. A Generic Cavity Array Problem

ended waveguides, horns and slots. This problem embodies all the same aspects as the general array problem except for the periodicity conditions on the unit cell side walls.

Figure 8 is a further simplification, which will be referred to as the "RF device problem." The interior region is again a cavity with perfectly conducting walls, as in Figure 7. But here, both inlet and outlet apertures lead into waveguides. This embodies all of the aspects of the cavity array problem except the periodic integral equation.

The RF device problem was the first stage of this research. It provided validation of the approach and implementation for the 3D finite elements and the waveguide aperture continuity conditions. A detailed summary of that work is given in a separate report [17]. The second research stage replaced the outlet waveguide modes with Floquet modes in order to solve the cavity array problem. The third and final stage in the algorithm and code development included the periodicity conditions needed for the general array problem. Each of the three solutions was validated by comparing computations to results published by other authors, obtained by methods other than FEM (measurements, mode matching, method of moments, etc.).

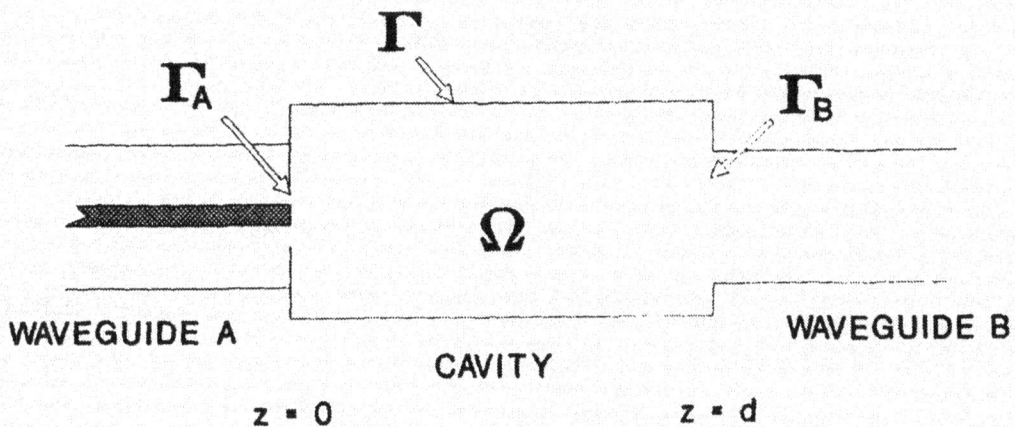

Figure 8. A Generic Passive RF Device Problem

III. Interior Region Problem - Finite Element Formulation

The first part of the problem formulation is the application of the finite element method to the interior region Ω. This will ignore, for the time being, the field continuity conditions on its enclosing boundary. This chapter will discuss the variational form of the problem and the restrictions it imposes on the electric field approximation. It then discusses the nature of the vector finite elements and shows that they satisfy the restrictions. Finally, it gives the derivation of the interior matrix terms using Galerkin's method, and their reduction to algebraic expressions using coordinate transformations local to each tetrahedron.

3.1. The Variational Statement

The boundary value problem consists of the operator equation (the vector wave equation), boundary conditions and applied forces. The solution is a function, the electric field, defined throughout Ω. The finite element method attempts to solve a variational equivalent of the problem.

The *variational statement* consists of a functional (usually an integral containing the unknown function in the integrand) and *admissibility* restrictions on the function [18]. Admissible functions are those that are in the domain of the functional and satisfy the boundary conditions. Appendix A discusses the two forms of functionals commonly used for vector electromagnetic problems and gives the rational for selecting the weak form (3).

There are three admissibility restrictions. First, the divergence condition $\nabla \cdot (\epsilon \bar{E}) = 0$ is necessary to ensure a unique solution since the operator equation only specifies the curl of $\bar{E}$. Second, the tangential electric field must vanish at the surface of perfect conductors, i.e. $\hat{n} \times \bar{E} = 0$. Last, at interfaces between dielectrics, tangential $\bar{E}$ is continuous, but normal $\bar{E}$ is discontinuous, i.e. $\hat{n} \cdot (\epsilon_1 \bar{E}_1) = \hat{n} \cdot (\epsilon_2 \bar{E}_2)$. It will be shown that these three restrictions are

16

satisfied through a careful choice of the expansion functions used to approximate $\bar{E}$.

3.2. Scalar vs. Vector Finite Elements

The most conventional finite elements are linear functions defined relative to the mesh nodes. Within a single tetrahedron there are four such functions, one per node. Each finite element is defined within a single tetrahedron and is zero everywhere else. Scalar node-based expansion functions for $\bar{E}$ are assembled from the finite elements. For example, in Figure 9, there are eight tetrahedra surrounding the central node. The scalar expansion function is defined over all eight cells, is equal to 1 at the center node, and goes linearly to zero at all surrounding nodes. In order to represent a vector field, one choice is to expand it in terms of these scalar functions with vector coefficients:

$$\bar{E} \approx \sum_{s=1}^{M} \bar{e}_s \, \phi_s(x,y,z) \qquad (4)$$

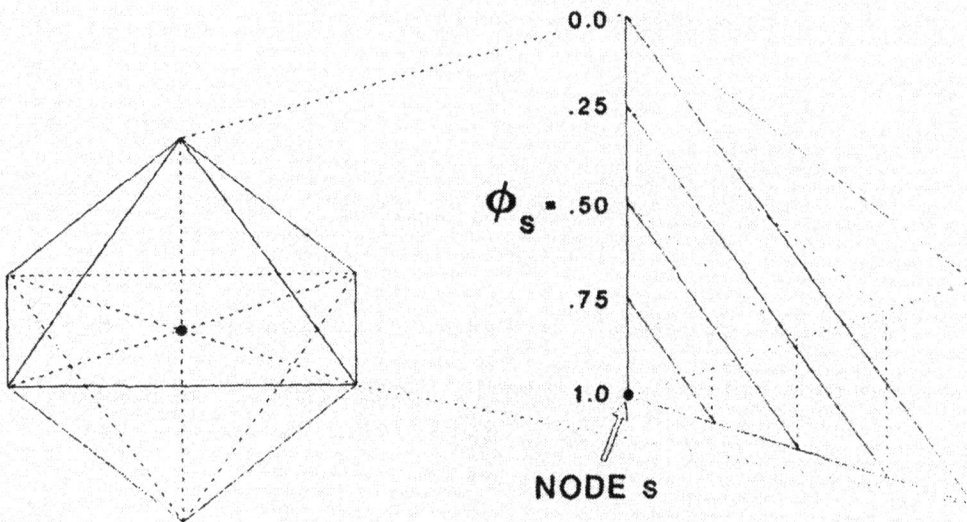

Figure 9. A Three-Dimensional Linear Expansion Function: (left) Eight Tetrahedra Surrounding a Node; (right) Linear Finite Element with Surfaces of Constant Function Value

17

where M is the total number of nodes in the mesh. However, this expansion has three important disadvantages:

(a) The boundary conditions at perfect conductors are difficult to enforce, especially at edges and tips where the surface normal is undefined.

(b) At dielectric interfaces where ϵ is discontinuous, the electric field normal to the interface is discontinuous, while the tangential components are continuous. But if a node s is on such an interface, (4) implies that all components are continuous. Hence, a node-based formulation will not accurately predict the field behavior at dielectric boundaries [19].

(c) It does not generally satisfy the divergence condition. Failure to enforce the condition will lead to spurious non-physical solutions [20]. It has been widely presumed that the penalty function method could be used, but Boyse et. al. point out that penalty methods are only justified for positive definite functionals [21], and (3) is indefinite.

Many of these difficulties can be circumvented by using vector finite elements in an expansion of the form

$$\bar{E} \approx \sum_{s=1}^{N} e_s \bar{\psi}_s(x,y,z) \tag{5}$$

where now s is an edge index and N is the number of edges in the mesh. The particular form of $\bar{\psi}$, sometimes attributed to Nedelec [22] that has been used most successfully is [23],[24]

$$\bar{\psi}_s = L_{ij}(f_i \nabla f_j - f_j \nabla f_i) \tag{6}$$

where f_i and f_j are the linear scalar finite elements defined for the nodes i and j bounding edge s. L_{ij} is the length of the edge, and is included as a scaling to ensure that the component of $\bar{\psi}$ tangential to the edge is a unit vector. Figure 10 illustrates the two dimensional version of this

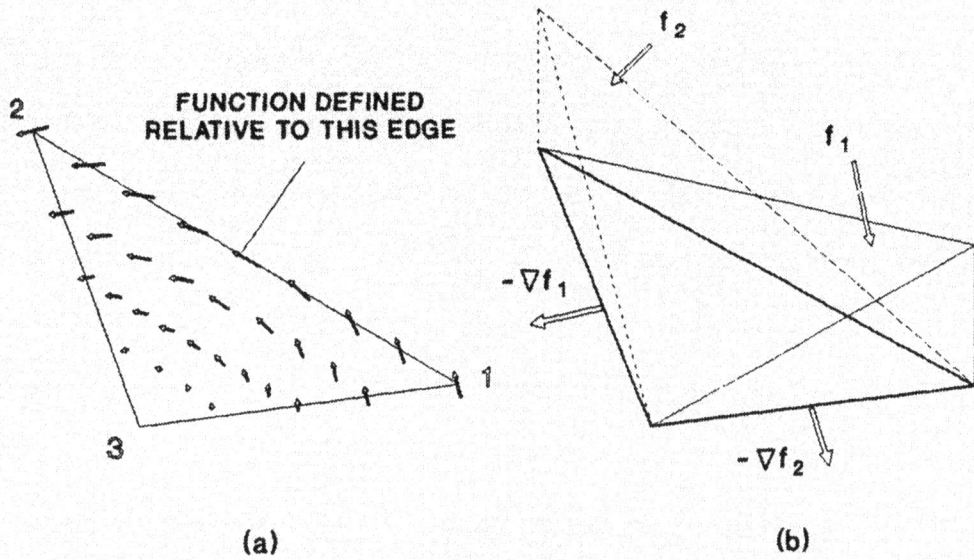

Figure 10. A Two-Dimensional Vector Finite Element: (a) Direction (arrow length) indicates function magnitude); (b) Constituent Linear Scalar Finite Elements

vector function and the scalar finite elements from which it is constructed. This choice resolves all three of the difficulties associated with the node based formulation since: (a) it allows a simple yet effective means for imposing conductor boundary conditions; (b) it enforces continuity of tangential field, but allows the normal field to be discontinuous at element boundaries; and (c) it is free of divergence since

$$\nabla \cdot \vec{V}_s = \nabla f_j \cdot \nabla f_i + f_i \nabla \cdot \nabla f_j - \nabla f_i \cdot \nabla f_j - f_j \nabla \cdot \nabla f_i$$

$$= f_i \nabla^2 f_j - f_j \nabla^2 f_i = 0$$

(7)

(f_i and f_j are linear, so their second derivatives are zero).

The principal disadvantages of the vector elements are algorithmic: (a) most CAD software generates a node listing that must now be converted to an edge listing; (b) the direction of each edge vector must be accounted for; and (c) the calculated fields must be converted back to vector components for output or display. One further disadvantage that has been cited [19],

19

[23] is that the number of unknowns is larger since there are typically 4-5 times as many edges as nodes in a tetrahedron mesh. Hence, there may be ⅓ - ⅔ more unknowns. This estimate is excessive for three reasons: First, it assumes three unknowns per node, but four are actually necessary to achieve an effective node-based formulation, using vector and scalar potentials [25]. Second, there are no unknowns associated with edges on perfect conductors, since the tangential electric field must be zero, whereas all three components of the electric field at a node on a conductor could be nonzero. Third, the node-based formulation may require finer sampling near conducting edges and corners to compensate for the uncertainty in the direction of n̂. Furthermore, the connectivity between edges is lower than for nodes, typically by a factor of 2, and hence the number of matrix entries is smaller by the same factor.

Given the above facts, the vector finite elements are clearly the better choice. The next section will show the derivation for the interior matrix terms using the expansion (5) in conjunction with Galerkin's method.

3.3. Discretization via Galerkin's Method

Galerkin's method is a specialization of weighted residuals, in which the trial functions are the same as the expansion functions. Its use is permitted when the expansion functions are in the admissible space of both the direct and the adjoint problem, as discussed in Appendix A.

Substitution of the series expansion for electric field into the operator equation leaves a *residual error* $\bar{R} = L(\bar{E}) - L(\hat{E})$ where $\hat{E}$ is the series approximation to $\bar{E}$ and L is the linear operator $[\nabla \times \mu_r^{-1} \nabla \times + k_0^2 \epsilon_r]$. The inner product of $\bar{R}$ with a trial function $\bar{W}$ is

$$\int_\Omega \bar{R} \cdot \bar{W}^* \, dv = F(\bar{E}) - \int_\Omega \sum_{i=1}^N e_i \left[\mu_r^{-1} \nabla \times \bar{\psi}_i \cdot \nabla \times \bar{W}^* \right.$$
$$\left. - k_0^2 \epsilon_r \bar{\psi}_i \cdot \bar{W}^* \right] dv + j k_0 \eta_0 \int_\Gamma \bar{J} \cdot \bar{W}^* \, ds \tag{8}$$

This includes the original functional since $\langle L(\bar{E}), \bar{W} \rangle = F(\bar{E})$, but since $F(\bar{E}) = 0$ from (3). The

20

weighted residuals procedure forces $\langle \bar{R}, \bar{W} \rangle = 0$ in order to solve for the coefficients e_t, giving

$$0 = \int_\Omega \sum_{t=1}^{N} e_t \left[\mu_r^{-1} \nabla \times \bar{\psi}_t \cdot \nabla \times \bar{W}^* - k_0^2 \epsilon_r \bar{\psi}_t \cdot \bar{W}^* \right] dv + j k_0 \eta_0 \int_\Gamma \bar{J} \cdot \bar{W}^* ds \qquad (9)$$

Substituting each $\bar{\psi}_s$, one at a time, for $\bar{W}$ gives N equations:

$$\int_\Omega \sum_{t=1}^{N} e_t \left[\mu_r^{-1} \nabla \times \bar{\psi}_t \cdot \nabla \times \bar{\psi}_s - k_0^2 \epsilon_r \bar{\psi}_t \cdot \bar{\psi}_s \right] dv + j k_0 \eta_0 \int_\Gamma \bar{J} \cdot \bar{\psi}_s ds = 0 \qquad (10)$$

The order of summation and integration in (10) may be reversed since the coefficients e_t are finite and the functions $\bar{\psi}_t$ and $\nabla \times \bar{\psi}_t$ are bounded. Then (10 defines a system of N equations in the N unknowns e_t. The volume integral terms are the entries in the matrix S^{EE} from (1):

$$S_{st}^{EE} = \int_{\Omega_{st}} \left[\mu_r^{-1} \nabla \times \bar{\psi}_s \cdot \nabla \times \bar{\psi}_t - k_0^2 \epsilon_r \bar{\psi}_s \cdot \bar{\psi}_t \right] dv \qquad (11)$$

The expansion and testing functions are each defined only on the collection of tetrahedra adjacent to the corresponding mesh edges. Hence the integration is over Ω_{st}, the collection of cells shared by edges s and t. These matrix equation terms will be computed by carrying out the integrations in (11) analytically using a transformation to homogeneous coordinates.

3.4. Homogeneous Coordinates

The *homogeneous* or *simplex* coordinates are defined locally within each tetrahedron [26:266-274]. There are four coordinates t_1, t_2, t_3 and t_4, but one of the four can always be eliminated using the relationship $t_1 + t_2 + t_3 + t_4 = 1$. The coordinate t_i of a point anywhere within the cell is the distance to node i from the opposing face, normalized to the cell height along that direction. Hence t_i is equal to one at node i; and zero at all other nodes as well as everywhere on the opposing face. The transformation is given in terms of a 4x4 matrix [T], whose elements are the 16 cofactors of the following matrix, U, made up of the cell vertex coordinates:

21

$$U = \begin{bmatrix} 1 & x_1 & y_1 & z_1 \\ 1 & x_2 & y_2 & z_2 \\ 1 & x_3 & y_3 & z_3 \\ 1 & x_4 & y_4 & z_4 \end{bmatrix} \qquad (12)$$

For example, $T_{22} = (y_3 z_4 - y_4 z_3) + y_1(z_3 - z_4) + z_1(y_4 - y_3)$. The four homogeneous coordinates are given as follows in terms of x,y,z:

$$\begin{bmatrix} t_1 \\ t_2 \\ t_3 \\ t_4 \end{bmatrix} = \frac{[T]}{6V} \begin{bmatrix} 1 \\ x \\ y \\ z \end{bmatrix} \qquad (13)$$

where V is the tetrahedron volume. These coordinates are especially convenient since the scalar finite elements become functions of one coordinate only:

$$f_i(x,y,z) = t_i = \frac{1}{6V}\left[T_{i1} + x T_{i2} + y T_{i3} + z T_{i4} \right] \qquad (14)$$

$$\nabla f_i = \frac{1}{6V}\left[\hat{x} T_{i2} + \hat{y} T_{i3} + \hat{z} T_{i4} \right] \qquad (15)$$

and the limits of integration are simplified. Most terms of (11) will reduce to integrals of products of two scalar functions, which have the simple result:

$$\iiint\limits_{cell} f_i f_j \, dx \, dy \, dz = 6V \int_0^1 dt_1 \int_0^{1-t_1} dt_2 \int_0^{1-t_1-t_2} t_i t_j \, dt_3 = \frac{V}{20}\left(1 + \delta_{ij} \right) \qquad (16)$$

where δ_{ij} is the Kronecker delta and i and j may take on any values between 1 and 4.

3.5. Volume Integral Computations

The volume integral computations are carried out by visiting each cell once and adding

22

its contribution to S_{st}^{EE} for every pair of edges that are part of the cell, excluding those edges that are on perfect conductors. Let i,j and m,n be the node indices of the endpoints of edges s and t, respectively, $1 \leq i,j,m,n \leq 4$, $i \neq j$, $m \neq n$. s and t are global indices, but i,j,m and n are local indices defined relative to a cell. Using the identity $\nabla \times (a\nabla b) = a\nabla \times \nabla b + \nabla a \times \nabla b$ and noting that the second derivatives of the linear functions f are all zero,

$$\nabla \times \vec{\psi}_s = 2L_s \nabla f_i \times \nabla f_j \tag{17}$$

$$\nabla \times \vec{\psi}_t = 2L_t \nabla f_m \times \nabla f_n \tag{18}$$

Considering the first term of (11) separately, note that the gradient terms (17) and (18) are constants and may be taken outside the integral. Thus, cell k's contribution to the first volume integral is

$$
\begin{aligned}
S_{st(k)}^{EE1} &= \frac{4V_k}{\mu_r} L_s L_t \nabla f_i \times \nabla f_j \cdot \nabla f_m \times \nabla f_n \\
&= \frac{4V_k L_s L_t}{\mu_r (6V_k)^4} \big[(T_{i3}T_{j4} - T_{i4}T_{j3})(T_{m3}T_{n4} - T_{m4}T_{n3}) \\
&\quad + (T_{i4}T_{j2} - T_{i2}T_{j4})(T_{m4}T_{n2} - T_{m2}T_{n4}) \\
&\quad + (T_{i2}T_{j3} - T_{i3}T_{j2})(T_{m2}T_{n3} - T_{m3}T_{n2}) \big]
\end{aligned}
\tag{19}
$$

where v_k is the volume of cell k. The cell's contribution to the second volume integral is

$$
S_{st(k)}^{EE2} = -k_0^2 \epsilon_r L_s L_t \int_{\Omega_k} \big[f_i f_m \nabla f_j \cdot \nabla f_n - f_j f_m \nabla f_i \cdot \nabla f_n \\
- f_i f_n \nabla f_j \cdot \nabla f_m + f_j f_n \nabla f_i \cdot \nabla f_m \big] dv
\tag{20}
$$

Again, the gradient terms are constants, so this may be evaluated using (16). The result is

23

$$S_{st(k)}^{EE2} = -\frac{k_0^2 \epsilon_r L_s L_t}{720 V_k} \sum_{l=2}^{4} \left[(1+\delta_{im}) T_{jl} T_{nl} - (1+\delta_{jm}) T_{il} T_{nl} \right.$$
$$\left. - (1+\delta_{in}) T_{jl} T_{ml} + (1+\delta_{jn}) T_{il} T_{ml} \right] \tag{21}$$

The submatrix S^{EE} in (1) is the sum of S^{EE1} and S^{EE2}. Equations (19) and (21) are a closed-form evaluation of the volume integral (11), written as an algebraic expression in terms of the geometry of a cell and the constitutive parameters contained within it.

This completes the discussion of the interior finite element solution. The next two chapters discuss the integral equations for the regions exterior to Ω. Those integral equations will provide expressions for $\bar{J}$ in terms of the transverse $\bar{E}$ on the boundary.

24

IV. Waveguide Continuity Conditions

This chapter develops the integral equations for the RF device problem illustrated in Figure 8. These equations apply to the two waveguides entering and leaving the cavity region, and are in the form of sums over waveguide modes. The derivation presented here is generic, applying to any type of waveguide for which eigenmode solutions are available. Appendix B gives specialized expressions for rectangular, circular and circular coaxial waveguides. Using the mode sum form, the computation of matrix entries will be shown to reduce to inner products of the mode functions with vector finite elements.

4.1. Combined-Source Integral Equation and Modal Expansion

The derivation of the integral equation uses the equivalence model shown in Figure 11. The interior problem sees zero field outside the cavity region and equivalent electric and magnetic currents on the open aperture. Those apertures are assumed to be planar and located in the end walls of the two waveguides. They do not necessarily extend across the entire waveguide cross-section (irises are allowed). The exterior problem sees zero field inside the cavity and oppositely-directed equivalent currents. Notice that when the interior and exterior problems are superposed, the equivalent currents cancel and the original problem is recovered. The integral equation applies to the exterior problem: It gives the fields in the waveguides in terms of the equivalent currents. The following approach is similar to that used by Harrington & Mautz in a moment method solution for open-ended waveguide radiation [27].

Let the field in waveguide A ($z < 0$) be comprised of a unit-amplitude incident field in the dominant mode traveling in the $+z$ direction, plus a series of reflected modes traveling in the $-z$ direction. The transverse (to z) electric field may be expressed in terms of a complete set of orthonormal mode functions $\bar{g}_i$ [28]:

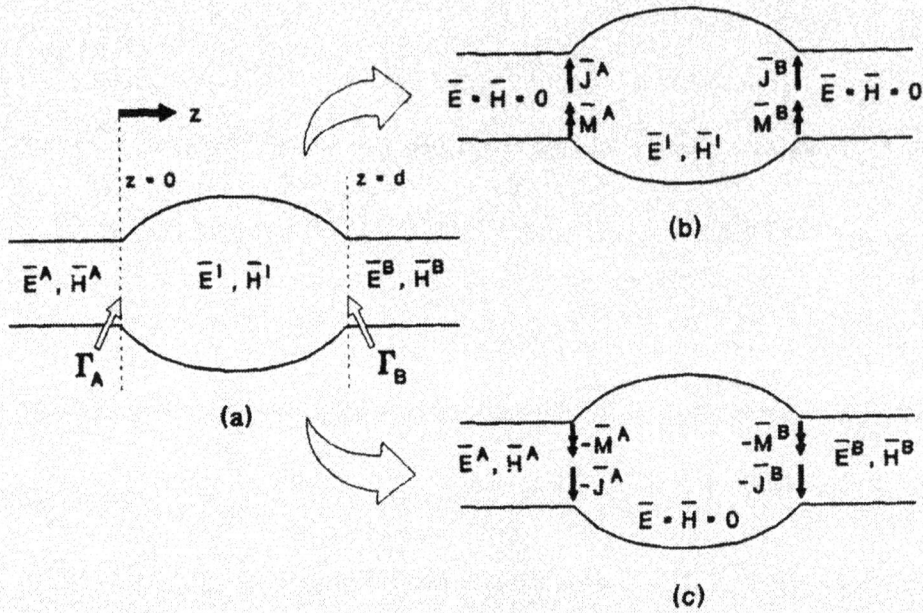

Figure 11. Equivalence Model for Waveguide/Cavity Problem: (a) Original Problem; (b) Interior Equivalent; (c) Exterior Equivalent

$$\bar{E}_t^A = e^{-\gamma_0 z}\,\bar{g}_0 + \sum_{i=0}^{\infty} C_i\, e^{\gamma_i z}\,\bar{g}_i \tag{22}$$

The subscript t denotes transverse and the γ_i's are the propagation constants. The dominant mode function is $\bar{g}_0$. The sum over modes i includes both TE and TM, as well as both $\sin(\phi)$ and $\cos(\phi)$ degeneracies for circular and circular coaxial waveguides. The complex coefficients C_i are unknowns that may be expressed in terms of the solution for the transverse fields in the apertures, e.g.

$$C_i = \int_{\Gamma_A} \bar{E}_t^A \Big|_{z=0} \cdot \bar{g}_i\, ds - \delta_{0i} \tag{23}$$

where δ_{0i} is the Kronecker delta. The propagation constant for mode i is related to its cutoff wavenumber, k_{ci}, by

26

$$\gamma_i \approx \sqrt{k_c^2 - k_0^2} \tag{24}$$

which is positive imaginary for propagating modes and positive real for evanescent modes. The transverse magnetic field is

$$\bar{H}_t^A = Y_0 e^{-\gamma_0 z}(\hat{z} \times \bar{g}_0) - \sum_{i=0}^{\infty} C_i Y_i e^{\gamma_i z}(\hat{z} \times \bar{g}_i) \tag{25}$$

where Y_i is a modal admittance:

$$Y_i = \begin{cases} \dfrac{-j\gamma_i}{k_0 \eta_0 \mu_r} & \text{(TE)} \\[2mm] \dfrac{jk_0 \epsilon_r}{\gamma_i \eta_0} & \text{(TM)} \\[2mm] \eta^{-1} & \text{(TEM)} \end{cases} \tag{26}$$

The boundary condition at $z=0$ is $-\bar{J}^A = \hat{n} \times \bar{H}^A$ giving, from (25):

$$\bar{J}^A = \hat{z} \times \bar{H}_t^A \big|_{z=0} = -Y_0 \bar{g}_0 + \sum_{i=0}^{\infty} C_i Y_i \bar{g}_i \tag{27}$$

Substituting (23) gives the final form of the integral equation for waveguide A:

$$\sum_{i=0}^{\infty} Y_i \bar{g}_i \int_{\Gamma_A} \bar{E}_t^A \big|_{z=0} \cdot \bar{g}_i \, ds - \bar{J}^A = 2Y_0 \bar{g}_0 \tag{28}$$

Notice that the equivalent magnetic current is involved indirectly by $\bar{E}_t^A(z=0) = \bar{M}^A \times \hat{z}$. Similarly, the integral equation for waveguide B is

$$\sum_{i=0}^{\infty} Y_i' \bar{g}_i' \int_{\Gamma_B} \bar{E}_t^B \big|_{z=d} \cdot \bar{g}_i' \, ds - \bar{J}^B = 0 \tag{29}$$

The primes signify the fact that waveguide B may be a different type or size than waveguide A.

so it may have different mode functions and modal admittances. The right hand side is zero because there are no sources in wavegude B.

A general requirement that ensures uniqueness in aperture problems is that continuity of both transverse $\bar{E}$ and transverse $\bar{H}$ must be enforced across each aperture. This usually implies a requirement to solve for both transverse field components independently (or, alternatively, for both $\bar{M}$ and $\bar{J}$). However, the nature of the waveguide mode expansion links the transverse field components through modal admittances, that is, $\bar{E}_t$ and $\bar{H}_t$ are not independent. Therefore, it is only necessary to solve for one of the two, and the obvious choice is $\bar{E}$, for consistency with the interior solution.

4.2. Discretization

The integral equations (28) and (29) give expressions for $\bar{J}$ that are substituted into the boundary integral of (3). Hence, they are tested using the same trial functions, $\bar{\psi}_s$ as the interior volume integral term. For compactness of notation, let Ψ_{si}^A and Ψ_{si}^B denote the following inner products:

$$\Psi_{si}^A = \int\limits_{\Gamma_A} \bar{\psi}_s \cdot \bar{g}_i \, ds \tag{30}$$

$$\Psi_{si}^B = \int\limits_{\Gamma_B} \bar{\psi}_s \cdot \bar{g}_i \, ds \tag{31}$$

Appendix B discusses the methods by which these integrals are computed for rectangular, circular and circular coaxial mode functions. The testing procedure gives the equations

$$E_s^{inc} = 2jk_0\eta_0 \, Y_0 \, \Psi_{s0}^A. \; s \in \Gamma_A \tag{32}$$

28

$$S_{st}^{EJ} = jk_0\eta_0 \sum_{i=0}^{\infty} Y_i \Psi_{si}^A \Psi_{ti}^A , \ s,t \in \Gamma_A \qquad (33)$$

$$S_{st}^{EJ} = jk_0\eta_0 \sum_{i=0}^{\infty} Y_i' \Psi_{si}^B \Psi_{ti}^B , \ s,t \in \Gamma_B \qquad (34)$$

In practice, the mode sums may be truncated at 32 or less, depending on: (a) the waveguide type; (b) the nature of the obstruction at and near the aperture; and (c) the ratio of the wavenumber to the cutoff wavenumber (more modes required near cutoff) [17]. Note that there is an entry S_{st}^{EJ} for every pair of edges s and t that share the same aperture, regardless of whether or not they share any mesh cells. Hence, this matrix is not sparse. E^{inc} has terms for all edges s that are in aperture A.

4.3. S Parameters

The performance of passive RF devices is typically expressed in terms of their "scattering," or "S" parameters. They may be found from those coefficients of the solution vector E that correspond to mesh edges in the two apertures. The modal excitation coefficients may be evaluated from these as

$$C_i = \sum_{s \in \Gamma_A} e_s \Psi_{si}^A - \delta_{0i} \qquad (35)$$

$$C_i' = \sum_{s \in \Gamma_B} e_s \Psi_{si}^B \qquad (36)$$

The coefficient C_0 is the reflection coefficient, or S_{11}. The transmission coefficient into each mode of waveguide B is

$$\tau_i = C_i' \sqrt{Y_i' / Y_0} \qquad (37)$$

When there is only one propagating mode in waveguide B, $\tau_0 = S_{21}$. If there is also only one propagating mode in waveguide A, then the following conservation of power relationship must hold: $|S_{11}|^2 + |S_{21}|^2 = 1$.

The derivations of this chapter, combined with those of Chapter III, provide a framework for a computer solution for two-port RF device S parameters. The computer code implementation and validation results are presented later in Chapter VII. The next chapter discusses how this methodology is extended to solve for the properties of phased arrays of cavity radiators.

V. Periodic Radiation Condition

The cavity array problem (Figure 7) was formulated as a straightforward extension of the RF device problem (Figure 8). It required replacing the integral equation for the second waveguide with one appropriate to a radiating aperture in an infinite array. This chapter gives the form of the integral equation for an infinite periodic array and shows how it is reduced to matrix form.

5.1. Periodic Integral Equation

The equivalence model for the cavity array problem is essentially the same as Figure 11. However, the aperture on the right side, formerly waveguide B, is now one aperture in an infinite uniform lattice. Therefore, the equivalent currents extend indefinitely over the outlet aperture plane in both x and y directions.

Each radiator in the array is assumed to be excited by a unit-amplitude incident field in the waveguide from $z < 0$, but the excitation phase may be different for each element in order to produce a beam directed towards angles θ_0, ϕ_0 in spherical coordinates. The phase shift as a function of x and y is

$$\phi(x,y) = e^{-j\psi_x x} \, e^{-j\psi_y y} \tag{38}$$

$$\psi_x = k \sin\theta_0 \cos\phi_0 \tag{39}$$

$$\psi_y = k \sin\theta_0 \sin\phi_0 \tag{40}$$

Figure 12 illustrates the notation convention for an array lattice with an arbitrary skew angle γ. The aperture shape is arbitrary. The fields and equivalent currents in each aperture must have the same magnitude as a function of x and y. A mathematical statement of the phase relationship in (38) is Floquet's theorem [29]:

31

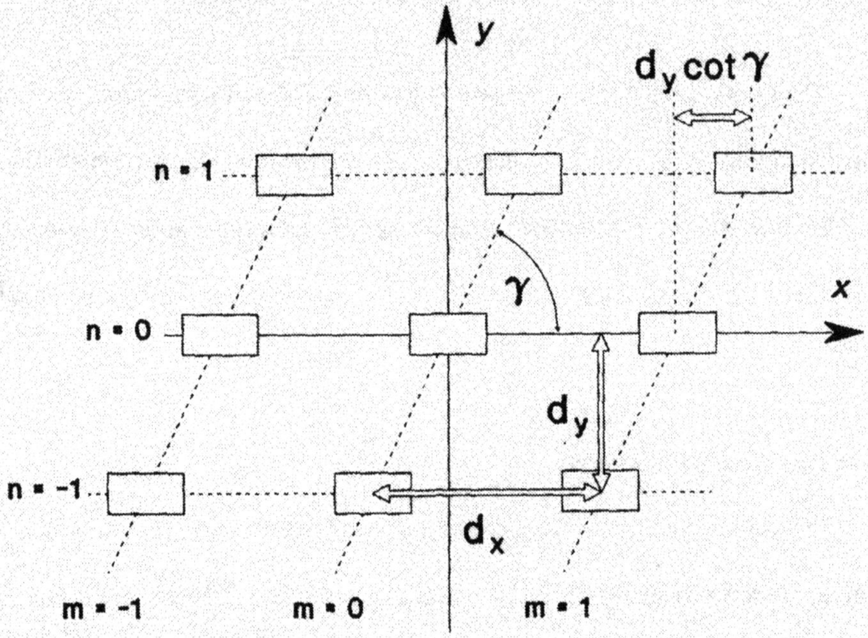

Figure 12. Infinite Array of Apertures, Skewed Lattice

$$\bar{J}(x+md_x+nd_y\cot\gamma, y+nd_y) = \bar{J}(x,y)\,e^{-j\psi_x(md_x+nd_y\cot\gamma)}\,e^{-j\psi_y nd_y} \qquad (41)$$

The left side is the equivalent current in any unit cell, and the right side is the current in the unit cell centered at the origin.

Equivalent currents $\bar{J}$ and $\bar{M}$ in the apertures will generate vector potentials $\bar{A}$ and $\bar{F}$ in the half space $z > d$. The magnetic field due to these is

$$\bar{H}(\bar{r}) = \nabla\times\bar{A} - j\omega\,\bar{F} + \frac{\nabla\nabla\cdot\bar{F}}{j\omega\mu\epsilon} \qquad (42)$$

The integral equation results from evaluating $\bar{H}_t$, the transverse magnetic field in the plane $z=d$ and using the boundary condition $\bar{J}=-\hat{z}\times\bar{H}_t$. Since $\bar{A}(z=d)$ is entirely z-directed it does not contribute to $\bar{H}_t(z=d)$ and the integral equation is

32

$$j\omega \hat{z} \times \left[\bar{F} + \frac{1}{k^2} \nabla\nabla \cdot \bar{F} \right] - \bar{J} = 0 \qquad (43)$$

$\bar{F}$ is an integral over the source current $\bar{M}$, and the limits of integration are infinite in x and y. Appendix C describes how it may be transformed into the following infinite summation:

$$0 = -\bar{J}(x,y) + \sum_{m=-\infty}^{\infty} \sum_{n=-\infty}^{\infty} \bar{\bar{T}}_{mn} \cdot \bar{\bar{E}}_{uc}(k_{xmn}, k_{ymn}) e^{-jk_{xmn}x} e^{-jk_{ymn}y} \qquad (44)$$

$$\bar{\bar{T}}_{mn} = \frac{\left[k^2 - k_{xmn}^2 - k_{ymn}^2 \right]^{-1/2}}{2 d_x d_y k \eta} \begin{bmatrix} (k^2 - k_{ymn}^2) & k_{xmn} k_{ymn} \\ k_{xmn} k_{ymn} & (k^2 - k_{xmn}^2) \end{bmatrix} \qquad (45)$$

$\bar{\bar{E}}_{uc}$ is the 2D Fourier transform of the transverse unit cell aperture field and k_{xmn} and k_{ymn} are sample points in the spatial frequency domain:

$$k_{xmn} = \frac{2\pi m}{d_x} - k_0 \sin\theta_0 \cos\phi_0 \qquad (46)$$

$$k_{ymn} = \frac{2\pi n}{d_y} - \frac{2\pi m \cot\gamma}{d_x} - k_0 \sin\theta_0 \sin\phi_0 \qquad (47)$$

sometimes referred to as *Floquet harmonics*. The summation in (44) may be computed numerically because its terms decay with increasing $|m|$ and $|n|$, as discussed further in Section 5.3.

5.2. Discretization

The integral equation (44) is reduced to matrix form using the procedure outlined in the previous chapter: First, solve for $\bar{J}$ and substitute it into the boundary term of (3); second, substitute the series expansion for $\bar{E}$; and third, substitute each $\bar{\psi}_s$ in turn for $\bar{W}$. Note that the integral equation involves the Fourier transform of the transverse aperture electric field, so its expansion will be in terms of the Fourier transforms of the $\bar{\psi}_t$'s:

33

$$\bar{E}_{uc} = \sum_{t=1}^{N} e_t \, \bar{\xi}_t (k_x, k_y) \tag{48}$$

$$\bar{\xi}_t (k_x, k_y) = \int\limits_{-\infty}^{\infty}\!\!\int \bar{\psi}_t \big|_{\Gamma_R} e^{jk_x x} e^{jk_y y} \, dx \, dy \tag{49}$$

where Γ_R denotes the radiating aperture. Testing gives the matrix terms:

$$S_{st}^{EJ} = \sum_m \sum_n \bar{\bar{T}}_{mn} \cdot \bar{\xi}_{tmn} \int\limits_{\Gamma_R} \bar{\psi}_s \, e^{-jk_{xmn}x} e^{-jk_{ymn}y} \, ds \, , \, s,t \in \Gamma_R \tag{50}$$

where $\bar{\xi}_{tmn}$ denotes $\bar{\xi}_t(k_{xmn}, k_{ymn})$. The limits of integration may be extended to $\pm\infty$ since $\bar{\psi}_s$ is zero outside Γ_R. Then, since ψ_s is a real function, the first integral may be recognized as the complex conjugate of the Fourier transform of ψ_s. (The Fourier transform of a real function is Hermitian, i.e. $\underline{F}(-k)=\underline{F}^*(k)$ [30:193]). Hence, a final expression for the elements of the matrix S^{EJ} is

$$S_{st}^{EJ} = \sum_m \sum_n \bar{\xi}_t(k_{xmn}, k_{ymn}) \cdot \bar{\bar{T}}_{mn} \cdot \bar{\xi}_s^*(k_{xmn}, k_{ymn}) \, , \, s,t \in \Gamma_R \tag{51}$$

5.3. Floquet Mode Limits

The infinite sums in (51) must be truncated at some upper and lower limits $\pm m$ and $\pm n$. Those limits are easily determined from the form of $\bar{\xi}$, derived in section C.4. Figure 13 is a contour plot of $|\bar{\xi}|$ in dB for a typical finite element. The axes are $k_x h_x$ and $k_y h_y$, where h_x and h_y are the triangle (mesh cell) heights parallel to the x and y axes. This scaling ensures that the size of the contours in Figure 13 are independent of mesh cell size. The Floquet harmonics are superimposed as dots in the figure. From (46) and (47), their locations (for a rectangular lattice and broadside scan) are $k_x h_x = 2\pi m h_x / d_x$ and $k_y h_y = 2\pi m h_x / d_x$. When the array scans away

34

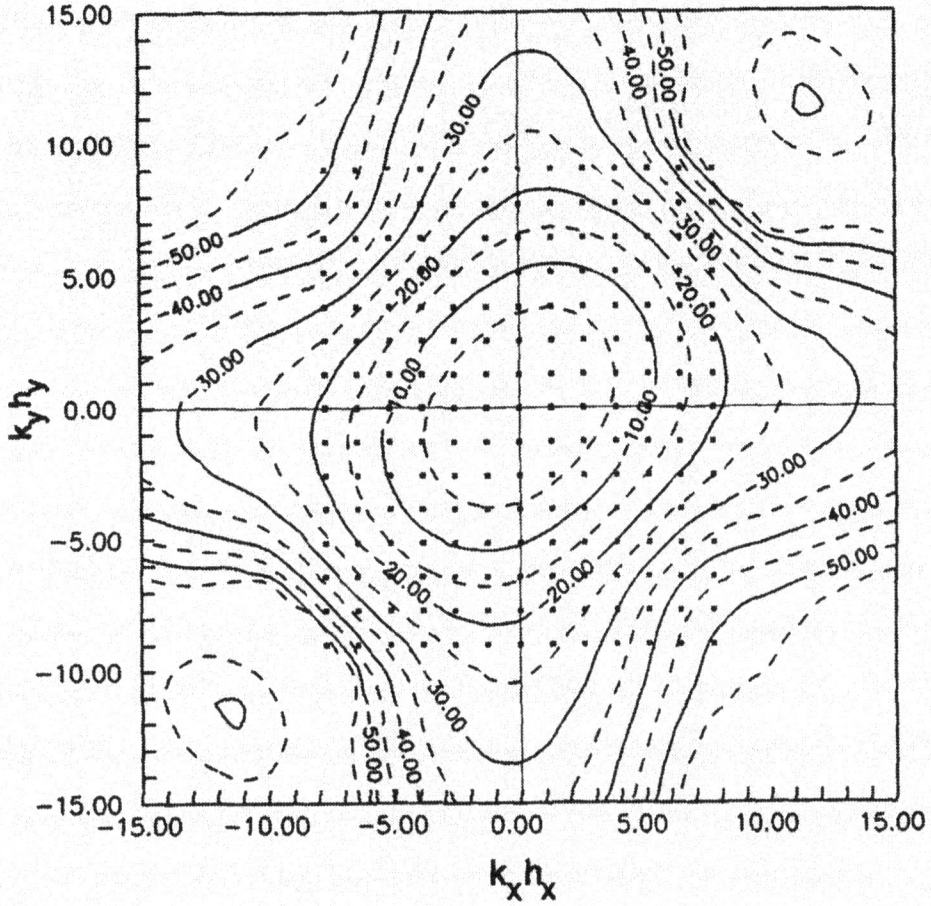

Figure 13. Typical Scalar Finite Element Fourier Transform (Contours in dB, sample points are Floquet modes for a square lattice with $h_x = h_y = d_x/5 = d_y/5$)

from broadside, the points will move, but their spacing will not change.

A reasonable upper limit on the number of sample points k_{xmn} and k_{ymn} that must be included in the computation of (51) are those inside the -20 dB contour. (The product of $\bar{\xi}_s$ and $\bar{\xi}_t$ will be less than -40 dB for any points outside that contour.) The size of the contour is consistently $k_x h_x \approx k_y h_y \approx \pm 2\pi$, but the sample spacing is inversely proportional to the unit cell lengths d_x and d_y. Hence, using $|k_x h_x|$, $|k_y h_y| \leq 2\pi$ in the leading terms of (46) and (47) gives $|m| \leq d_x/h_x$ and $|n| \leq d_y/h_y$. In a typical problem, d_x and d_y are each approximately

35

.5λ and the mesh cells are approximately .1λ in height. In such cases limits of $-5 \leq m,n \leq 5$ are adequate to ensure convergence. More modes must be included when the unit cell size is larger, or when the mesh cell size is smaller.

5.4. Active Reflection Coefficient and Element Radiation Pattern

Some of the most important output results from the phased array analysis are the active reflection coefficient, R_a, and the active element radiation (far field) pattern. Both are functions of scan angle. They are analogous to the reflection and transmission coefficients in the two port RF device problem.

The expression for R_a is identical to C_0 in (35). Both are due to the field reflected into the inlet waveguide, computed from the waveguide aperture field. But now those aperture fields include the effects of a periodic radiation condition at the outlet side, and so it is the reflection coefficient for one feed waveguide in an infinite array, i.e. the "active array reflection coefficient."

The active element pattern is analogous to a transmission coefficient. It is a measure of the excitation strength of a plane wave (a Floquet mode) propagating away from the array. Amitay et. al. show that the θ and ϕ polarization components of the element's far field pattern are due entirely to the lowest order TE and TM Floquet modes, respectively [29:57]. Rewriting their expressions in terms of Fourier transforms gives:

$$E_\theta = \frac{\sec\theta}{\sqrt{d_x d_y}} (\hat{x}\cos\phi + \hat{y}\sin\phi) \cdot \bar{\underline{E}}_{uc}^*(k_{x00}, k_{y00}) \tag{52}$$

$$E_\phi = \frac{1}{\sqrt{d_x d_y}} (\hat{y}\cos\phi - \hat{x}\sin\phi) \cdot \bar{\underline{E}}_{uc}^*(k_{x00}, k_{y00}) \tag{53}$$

A check for conservation of power may be made by computing the transmission coefficients for the two mn=00 Floquet modes, whose admittances are Y_{100} and Y_{200} for TM and TE (see section C.5):

$$T_\theta = E_\theta \sqrt{Y_{200}/Y_0} = E_\theta \sqrt{\sec\theta / Y_0 \eta_0} \qquad (54)$$

$$T_\phi = E_\phi \sqrt{Y_{100}/Y_0} = E_\phi \sqrt{\cos\theta / Y_0 \eta_0} \qquad (55)$$

Y_0 is the feed waveguide's dominant mode admittance. When the feed waveguide supports only one mode and there are no grating lobes in visible space, the conservation of power relationship is $|R_a| + |T_\theta| + |T_\phi| = 1$.

The derivations given in this chapter, when combined with the preceding chapters' finite element and waveguide derivations, constitute the framework for a computer solution for the scanning properties of cavity arrays. The implementation and validation results are discussed later in Chapter VII.

VI. Periodic Boundary Conditions

The final stage in the problem formulation bridges the gap between the cavity array problem in Figure 7 to the general array problem in Figure 4. The side walls of the cavity region Ω are no longer conductors, so adjacent radiators are free to interact across those walls. The periodic radiation condition does not account for that interaction. This chapter will show that the general array problem may be accomplished by constructing a matrix for a single unit cell as though it were a cavity with open-circuit side walls; then applying a mapping directly to the matrix to enforce the periodicity condition. The necessary characteristics for a unit cell will be discussed first, then the algorithm for the matrix mapping will be presented. Next, the side walls will be shown to have no net contribution to the boundary functional. The specialization of the algorithm to edges shared by the radiation boundary condition and a unit cell side wall is discussed last.

6.1. Unit Cell Representation

A typical radiator is fed by a waveguide through an aperture in a ground plane. Unlike the cavity radiators considered in the last chapter, it has some structure projecting above the ground plane. That structure may be enclosed by an imaginary box whose lower surface is the ground plane at $z=0$. Its top surface at $z=h$ is in free space above the radiator structure (see Figure 4). The side walls of Ω are the unit cell boundaries. As was indicated in Figure 5, the unit cell may be trapezoidal as well as rectangular. In fact, its shape is fairly arbitrary within a few constraints, with some possibilities illustrated in Figure 14. The constraints are: (a) the unit cell side walls do not cross the feed waveguide; and (b) opposing boundaries must be identical except for a translation of $(d_x,0)$ or $(d_y\cot\gamma,d_y)$. The first array (Figure 14a) has circular waveguides whose diameter is larger than d_y, so the unit cell shape along the boundaries has been

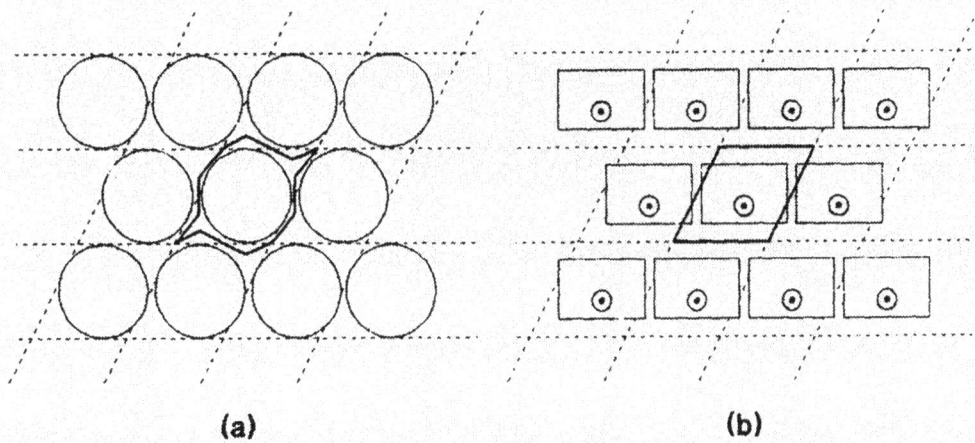

(a)   (b)

Figure 14. Unit Cell Definitions: (a) Circular Waveguide Array; (b) Rectangular Patch Array

altered. The second is a rectangular patch array, showing that the unit cell walls may cut through

the radiator's conducting structure.

The unit cell definition must be such that the Floquet condition is observed. Let $\bar{f}_{mn}$

denote a field in the m'th column and n'th row of the lattice. Then the unit cell fields are related

by:

$$\bar{f}_{m,n} = \bar{f}_{0,0} \; e^{-jm\alpha_x} \; e^{-jn\alpha_y}$$
$$\alpha_x = \psi_x d_x$$
$$\alpha_y = \psi_x d_y \cot\gamma + \psi_y d_y$$

(56)

where ψ_x and ψ_y are given by (39) and (40). An example unit cell mesh is shown in Figure 15.

The two perspective views show opposite sides of the mesh to illustrate the important requirement

that the surface mesh on opposing faces must be identical. Every edge on the +x or +y bound-

ary has an *image edge* on the -x or -y boundary respectively. Consequently, the expansion and

testing functions associated with those edges are identical except for a translation.

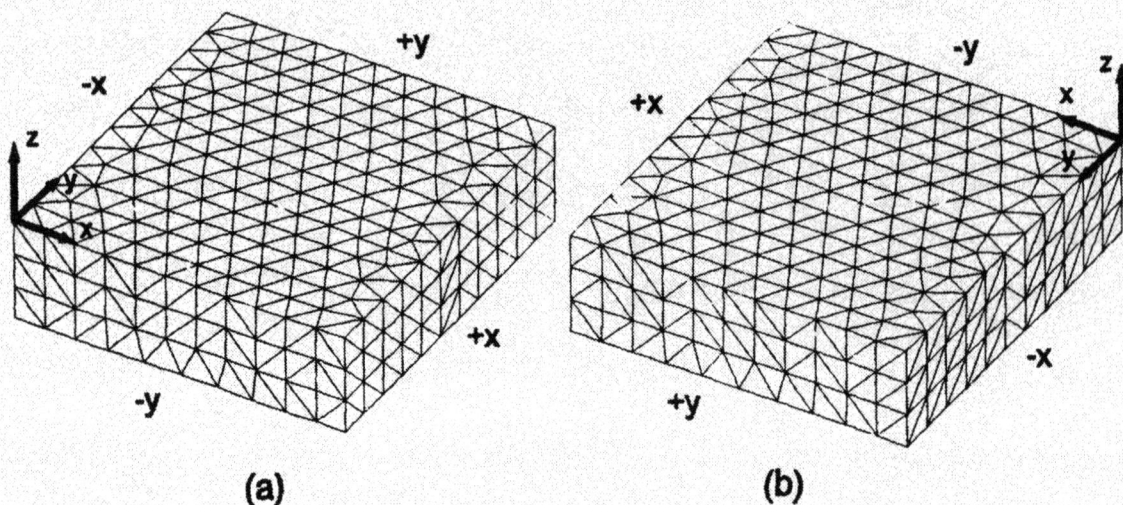

Figure 15. Example Unit Cell Mesh Showing Opposite Faces

6.2. *Mapping from an Infinite System*

The matrix generated by discretizing the functional represents interactions between electric field sources associated with mesh edges. Consider the 2D triangle mesh shown in Figure 16, which represents an infinite mesh of an infinite array. It was constructed in such a way that opposing unit cell boundaries have identical mesh edges. Consequently, when the mesh is replicated in each unit cell, the finite elements are perfectly aligned across unit cell boundaries.

If the finite element method were applied to the infinite problem, it would generate an infinite size matrix. Appendix D shows formally, for a one-dimensional problem, how periodic boundary conditions are exploited to reduce the problem to a finite matrix involving a single unit cell. The extension to periodicity in two dimensions is straightforward. Essentially, the procedure amounts to "folding" opposing unit cell boundaries onto each other. The +x and +y boundary edges are removed from the problem and the -x and -y boundary edges will interact with those just inside the opposing boundaries, but with an appropriate phase shift.

40

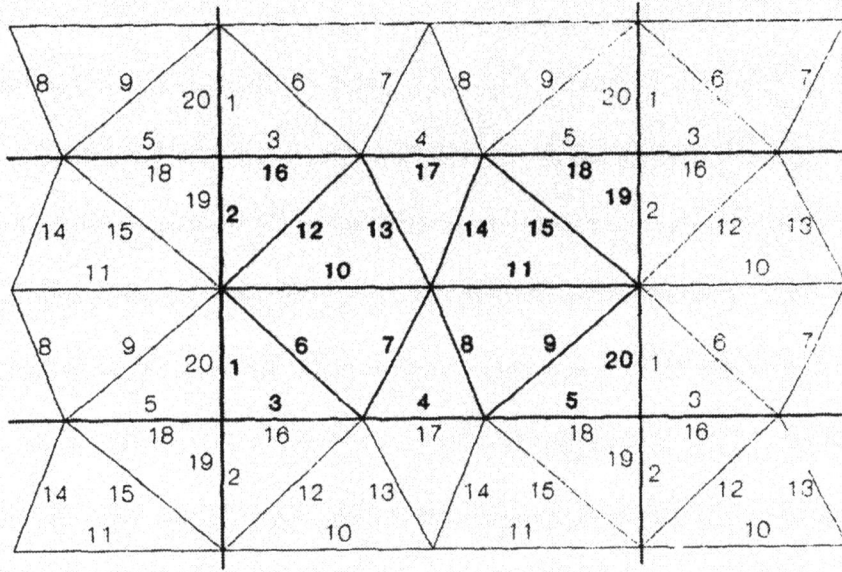

Figure 16. Triangular Finite Element Mesh for Infinite Two-Dimensional Periodic Problem

The first step in the procedure is to construct the matrix for an isolated unit cell. Next, the terms involving the +x boundary are removed, and corresponding -x boundary terms are created. Then the latter step is repeated for the +y and -y boundaries. For example, in Figure 16, edges 15 and 19 share a mesh cell (a triangle), so there will be matrix terms $S_{15,19}$ and $S_{19,15}$. Edge 2 is edge 19's image, so the new entries created are $S_{15,2}=S_{15,19}\exp\{-j\alpha_x\}$ and $S_{2,15}=S_{19,15}\exp\{j\alpha_x\}$. Also, $S_{19,19}$ will be added to $S_{2,2}$. This is necessary because of the truncation of the mesh at the unit cell boundary. In the infinite mesh, edge 2 would interact with itself through cells to the left and right. In the unit cell mesh, there is no cell to the left, so $S_{2,2}$ is incomplete, but $S_{19,19}$ is identical to the missing contribution. In the 3D problem with two-dimensional periodicity, there will also be matrix entries for edge pairs that are both on the +x or +y boundaries, representing their interaction within a cell to the left of the boundary. These terms must also be added to their -x counterparts without a phase shift.

41

The algorithm is summarized in Figure 17. A consequence of removing mesh edges is that some of the matrix rows and columns become entirely zero, and they must be deleted before the matrix solution can begin. This is consistent with the fact that unknowns associated with image edges are not independent. Due to periodicity, once one is known, the other follows from the periodicity condition. Unless one set of dependent unknowns is removed, the system would be over-determined.

6.3. Boundary Functional

The algorithm discussed above only dealt with the volume integral terms from the

I. FOR EVERY EDGE s ON +x BOUNDARY:

    A. LOCATE IMAGE EDGE s' ON -x BOUNDARY

    B. FOR EVERY EDGE t SUCH THAT $S_{st} \neq 0$:

        1. IF t IS ON THE + x BOUNDARY, THEN:

            a. LOCATE IMAGE EDGE t'

            b. SET $S_{s't'} = S_{s't'} + S_{st}$

            c. SET $S_{st} = 0$

        2. ELSE IF t IS NOT ON THE +x BOUNDARY, THEN:

            a. SET $S_{s't} = S_{st} \exp\{j\alpha_x\}$

            b. SET $S_{ts'} = S_{ts} \exp\{-j\alpha_x\}$

            c. SET $S_{st} = S_{ts} = 0$

II. REPEAT I FOR +y AND -y BOUNDARIES

III. COMPRESS THE MATRIX (ELIMINATE ZERO ROWS & COLUMNS)

IV. COMPRESS THE INCIDENT CURRENT VECTOR

Figure 17. Periodic Boundary Condition Algorithm

functional. The boundary integral must now account for the contributions from the unit cell side walls in addition to the waveguide aperture and radiation boundary. The boundary integral may be regarded as an expression for power flow across Γ. In an infinite array, the power entering one side of the unit cell must be the same as that leaving the opposite side, so it is expected that the side walls should have no net contribution. The following will show this to be true.

Consider the boundary integral of (3) at opposing unit cell walls parallel to the x-z plane. The magnetic field must obey the periodicity condition, and since the outward surface normals are opposite, the equivalent currents are

$$\bar{J}(x+2\beta,\frac{d_y}{2}) = -\bar{J}(x,-\frac{d_y}{2}) e^{-j\alpha_y}$$
$$\beta = \frac{1}{2} d_y \cot\gamma \tag{57}$$

Any admissible trial function $\bar{W}$ (see Appendix A) must obey the conjugate relationship because it represents waves traveling in the opposite direction, i.e.

$$\bar{W}^*(x+2\beta,\frac{d_y}{2}) = \bar{W}^*(x,-\frac{d_y}{2}) e^{j\alpha_y} \tag{58}$$

The boundary functional evaluated at the two unit cell walls is

$$F_{y-} = \int_0^h dz \int_{-\frac{d_x}{2}-\beta}^{\frac{d_x}{2}-\beta} \bar{W}^*(x,-\frac{d_y}{2}) \cdot \bar{J}(x,-\frac{d_y}{2}) \, dx \tag{59}$$

$$F_{y+} = \int_0^h dz \int_{-\frac{d_x}{2}+\beta}^{\frac{d_x}{2}+\beta} \bar{W}^*(x,\frac{d_y}{2}) \cdot \bar{J}(x,\frac{d_y}{2}) \, dx \tag{60}$$

Using a change of variables, let $\tau = x - 2\beta$:

$$F_{y+} = \int_0^h dz \int_{-\frac{d_x}{2}-\beta}^{\frac{d_x}{2}-\beta} \bar{W}^*(\tau+2\beta, \frac{d_y}{2}) \cdot \bar{J}(\tau+2\beta, \frac{d_y}{2}) \, d\tau \qquad (61)$$

Using (59) and (60):

$$F_{y+} = -\int_0^h dz \int_{-\frac{d_x}{2}-\beta}^{\frac{d_x}{2}-\beta} \bar{W}^*(\tau, -\frac{d_y}{2}) \cdot \bar{J}(\tau, -\frac{d_y}{2}) \, d\tau \qquad (62)$$

and therefore, $F_{y+} = -F_{y-}$. A similar procedure will show that $F_{x+} = -F_{x-}$, hence the unit cell side walls have zero net contribution. A similar result has been reported for a two-dimensional problem with periodicity in one dimension [16].

6.4. Radiation Boundary

A form of periodicity condition was already imposed on the radiation boundary through the periodic integral equation. However, it did not account for the mesh truncation at the unit cell boundary.

In order to complete the specification of periodicity conditions for edges on Γ_R, either of two approaches may be used: The first method is to apply the same algorithm (Figure 17) used for the finite element boundary terms, but excluding I.B.2. The second, demonstrated by Gedney for periodicity in one dimension [16] used "overlap basis functions" at one boundary edge. Those expansion and testing functions associated with points on one boundary extend into the next unit cell. In the context of Figure 16, that amounts to: (a) removing the expansion functions for edges on the +x and +y boundaries (16-20); and (b) extending the functions for edges 1 & 2 and 3-5 into the next unit cells to the left and below, respectively. Then the periodic

44

integral equation is applied to the modified mesh. Both techniques were tested as part of the code development and validation, giving equivalent results.

6.5. Summary

This chapter and the preceding three make up the formulation for the general phased array problem. They are the framework for mapping the boundary value problem into a discrete form that may be solved by computer. The matrix for the general array problem is constructed of the three parts that were developed in Chapters III, IV and V: first, a sparse matrix due to finite element interactions within the unit cell; second, a dense upper-left submatrix due to waveguide interactions; and third, a dense, lower-right submatrix due to Floquet-mode interactions. The side-wall periodicity conditions are implemented as a transformation of that matrix, as was described in this chapter. The next three chapters discuss the validation tests for three computer codes that implement the solutions to the RF device problem; the cavity array problem; and the general array problem.

VII. Validation - RF Device Problem

The goals for this simplified problem (Figure 8) were to demonstrate essential characteristics of the finite element and waveguide mode implementations. The 3D vector finite elements were shown to correctly predict the field behavior at conductor edges and dielectric interfaces. The higher-order waveguide mode calculations were validated through comparisons with measured and published results for several waveguide discontinuity problems. These results also provided estimates for sampling requirements: the number of finite elements; and the number of waveguide modes. More complete details are given in a previous report [17].

7.1. Computer Code Implementation

7.1.1. General Procedure. A FORTRAN computer code named TWOPORT implements the solution to this generic problem. An outline of the actions it takes is given in Figure 18. The user instructions, read during the first step, include: the type, size, location, number of modes and ϵ_r for each waveguide; the frequency limits and frequency stepsize; and the name of the file containing the problem geometry. The geometry file contains three sections: The first lists the node coordinates and several flags for each, identifying boundary nodes (port # for those in waveguide apertures and conductor # for those on conducting surfaces). The second block lists the indices of the four nodes comprising each tetrahedron and the index of the material filling it. The last block lists the complex ϵ_r and μ_r of each material.

The third step converts the node-based geometry to an edge-based geometry. Each edge in the mesh defines an electric field vector expansion function that exists over all cells adjacent to that edge. If two nodes are on the same conductor, then there cannot be a field along the line joining them, so those edges are not included in the edge list. This is the means of enforcing the boundary condition on tangential electric field at conducting surfaces. On the other hand, if two

46

```
┌─────────────────────────────────────────────────────────────────┐
│  I.  READ INSTRUCTIONS AND OPTIONS                                │
│                                                                   │
│  II.  READ PROBLEM GEOMETRY                                       │
│                                                                   │
│  III.  CREATE EDGE-BASED GEOMETRY                                 │
│                                                                   │
│  IV.  FOR EACH FREQUENCY:                                         │
│                                                                   │
│      A.  COMPUTE TERMS OF $S^{EE}$ ACCORDING TO (20), (22)        │
│                                                                   │
│      B.  FOR WAVEGUIDE A:                                         │
│                                                                   │
│          1.  COMPUTE INCIDENT CURRENT VECTOR IN (32)              │
│                                                                   │
│          2.  COMPUTE TERMS OF $S^{EJ}$ FROM (34), (35)            │
│                                                                   │
│      C.  FOR WAVEGUIDE B, COMPUTE TERMS OF $S^{EJ}$ FROM (34), (35)│
│                                                                   │
│      D.  SOLVE  $(S^{EE} + S^{EJ})\, E = E^{inc}$  FOR E          │
│                                                                   │
│      E.  COMPUTE MODE EXCITATION COEFFICIENTS FROM (36), (37)     │
└─────────────────────────────────────────────────────────────────┘
```

Figure 18. Solution Procedure in Program TWOPORT

nodes are on different conductors, there may be a field between them, and such edges must still be included. The matrix fill operations must account for the direction of the vector function, so, by convention, it is always directed from lowest to highest node index.

Most of the actions under step IV in Figure 18 are straightforward implementations of formulas derived in Chapters III and IV and Appendix B. The matrix elements are computed in three steps: first, the interior finite element interactions; and second and third, the exterior waveguide interactions for waveguides A and B, respectively. After solving the system for the unknown electric field coefficients, excitation coefficients for any number of higher order modes may be computed.

7.1.2. Matrix Solution. The matrix structure is mostly sparse, except for two dense submatrices in the upper left and lower right corners. This structure results from ordering the edges by increasing centroid z coordinate, starting with those in the waveguide A aperture and ending with those in the waveguide B aperture. Unfortunately, there is not usually any special structure (sparsity pattern) to the matrix. It tends to be banded, but the bandwidth is large and there is no advantage to using band storage or specialized band solvers. Hence the two approaches used for the matrix solution were: (a) ordinary LU decomposition (LUD) using standard library routines and ordinary row-column storage (IMSL [31:34] and LAPACK[32:150]) for problems with 2000 unknowns or less; and (b) the conjugate gradient method (CGM) based on formulas from Sarkar and Arvas [33], using sparse storage, for larger problems.

The CGM solver was written specifically to accommodate the form of sparse storage most attractive for this particular class of problems. The two dense submatrices due to the waveguides are stored in ordinary row, column format; while the sparse finite element matrix is stored in a column array. Its entries are in arbitrary order, stored in the order that each edge pair is first encountered as the fill algorithm performs operations one cell at a time. That assembly technique is considerably more efficient than performing the same operations one edge at a time [26].

Figure 19 is a comparison of execution times versus number of edges for LUD and CGM. The CGM solver clearly has the advantage for problems with 1000 unknowns or more. Unfortunately, its solution time is more difficult to predict *a priori*, since the number of iterations it will need to converge is unknown. The convergence measure is the residual error norm, which is the L_2 norm of the solution error:

$$\epsilon_i = \| S E_i - E^{inc} \| \tag{63}$$

(i is the iteration number). The initial guess E_0 is the zero vector. Figure 20 is an example convergence history (for a microstrip transmission line fed by coaxial waveguide at each end). The S parameters have converged to within 1% of their correct values when ϵ_i/ϵ_0 is less than .001. Problems with rectangular and circular waveguide ports typically only require 1N-2N itera-

48

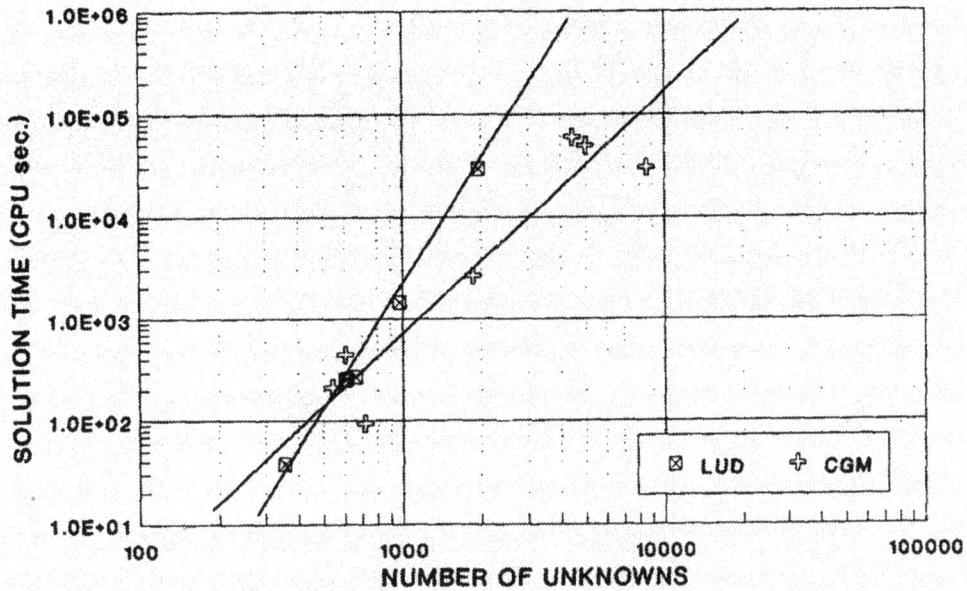

Figure 19. Comparison of Solution Times for LU Decomposition (IMSL) and Conjugate Gradient Method (CPU seconds on VAX® 8650 minicomputer)

tions to converge, where N is the number of unknowns.

7.2. Waveguide Discontinuities

7.2.1. Iris in Coaxial Waveguide. Two test cases are representative of the several waveguide discontinuity problems discussed in [17]: a conducting iris in coaxial waveguide; and a step discontinuity in circular waveguide. For the first of these, Figure 21 is the tetrahedron mesh used as the input to program TWOPORT. Relating this model to Figure 8, the inlet and outlet waveguides are both coaxial (inner radius a=1.5mm, outer radius b=3.5mm) and the cavity region Ω represented by the mesh is simply a short section of the same waveguide. Shading has been added to identify the nodes tagged as perfect conductors. One quadrant of the inlet end is blocked by a thin conducting iris.

Measurements of this "device" were made by inserting a foil iris between two APC-7mm adapters and using a network analyzer to obtain S_{11} over a 2-18 GHz frequency range. Figure

49

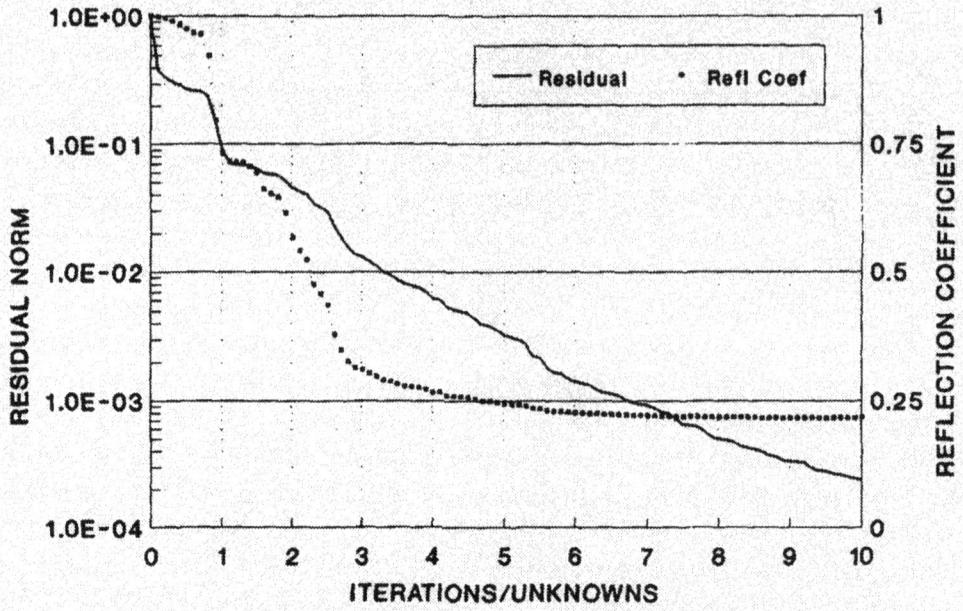

Figure 20. Convergence of Residual Norm and Reflection Coefficient
using Conjugate Gradient Matrix Solver

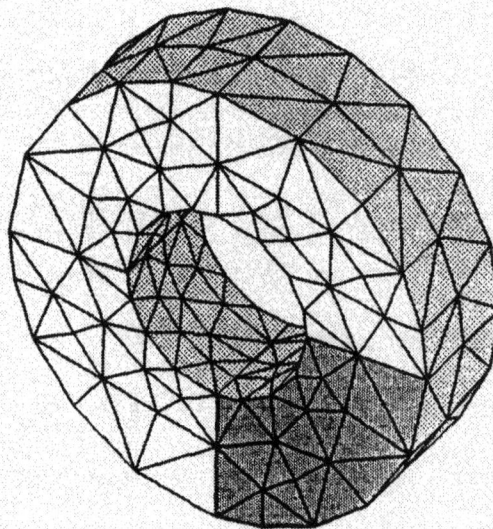

Figure 21. Finite Element Mesh for Coaxial Waveguide Section with Conducting Iris

50

22 is a comparison of those measurements with TWOPORT calculations, for both the magnitude and phase. These results validate two important features of the solution approach: First, the higher-order coaxial mode functions and their inner products with finite elements are correctly implemented. Second, the finite element solution is correctly predicting the field behavior at the edge of a perfect conductor.

7.2.2. Circular Waveguide Mode Converter. A step discontinuity in a circular wave-guide is a simple type of mode converter commonly used in feeds for reflector antennas [34]. The antenna's radiation pattern shape is controlled by carefully adjusting the amplitude ratio of modes in an oversize (multimode) waveguide or horn. Figure 23 shows finite element meshes for two test cases with different inlet waveguide radius and the same outlet waveguide radius. Multimode calculations for these geometries were presented by Masterman & Clarricoats [35]. Their results are in terms of a mode conversion ratio, M:

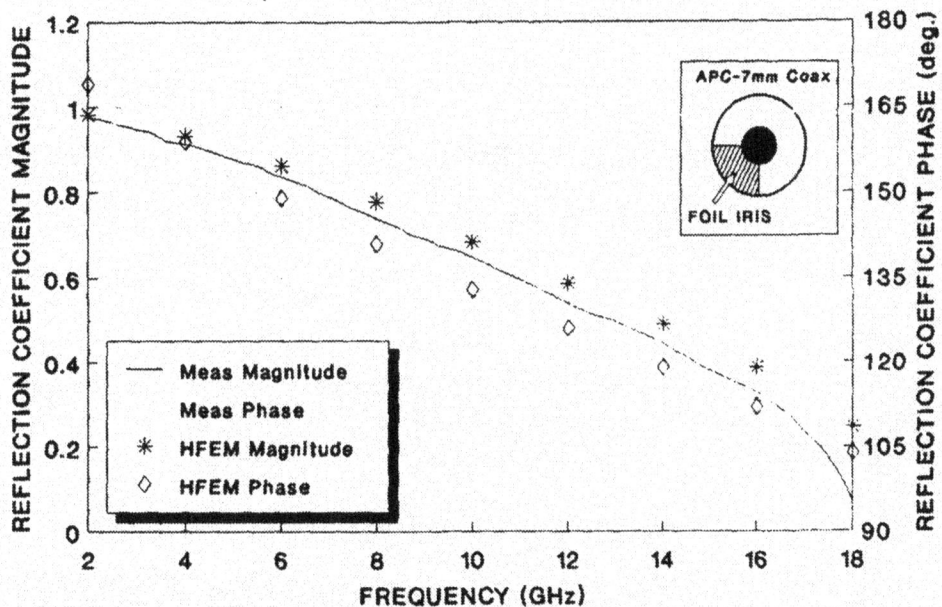

Figure 22. Measured and Computed S_{11} Magnitude and Phase for Coaxial Iris

$$M = \frac{|C_2' g_{\rho 2}'(\rho = a)|}{|C_1' g_{\rho 1}'(\rho = a)|} \qquad (64)$$

C_1' and C_2' are the excitation coefficients for the TE_{11} and TM_{11} modes, respectively, and $g_{\rho 1}'$ and $g_{\rho 2}'$ are the radial components of the mode functions. Hence, M is the relative strength of the two modes measured at the wall of the outlet waveguide.

Figure 24 compares TWOPORT calculations with the multimode results. The discontinuity at 6.25 GHz for $a_1 = 1.15''$ is due to the fact that the inlet waveguide also supports the TM_{11} mode above that frequency. The agreement of the TWOPORT calculations with the well-established multimode calculations indicates that the higher-order circular waveguide mode functions and their finite element inner products have been implemented correctly. It also demonstrates a flexibility of the approach of using higher-order modes in modeling feed structures that support more than one propagating mode.

7.3. Printed Circuit Devices

One of the most important capabilities of the finite element method is its ability to deal with inhomogeneous dielectrics. Printed circuit devices are an example application that is especially interesting to this research because their modeling requirements are similar to the printed antennas discussed in the introduction. The specific test cases were: (a) a straight length of microstripline with coaxial connectors; and (b) a microstrip meander line.

7.3.1. Microstrip Transmission Line. The finite element mesh shown in Figure 25 is a thin dielectric slab (100 μm), situated in the bottom of a perfectly conducting box. The mesh for the air space above the slab is not shown. Shading has been added to identify the nodes associated with microstrip line, the coax center conductor, and the coax dielectric. There is an identical coaxial port at the far end of the device. The coax dimensions (a = 43 μm, b = 100 μm) were

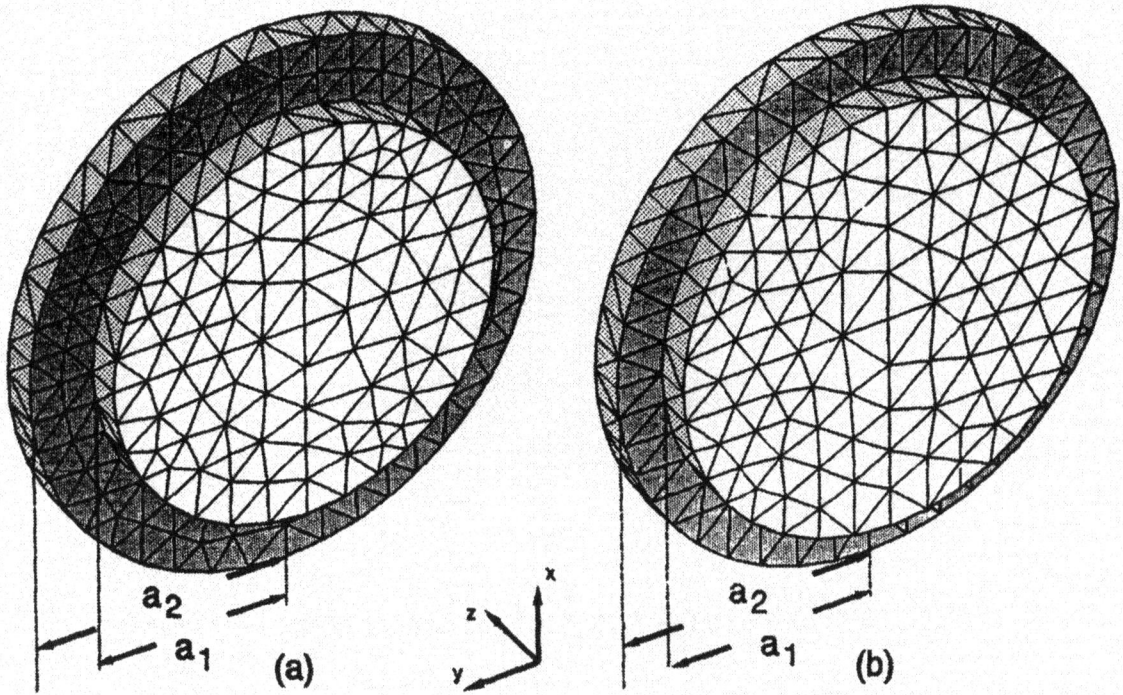

Figure 23. Finite Element Meshes for Circular Waveguide Step Discontinuities (Mode Converters) with $a_2 = 1.4"$: (a) $a_1 = 1.05"$; (b) $a_1 = 1.15"$.

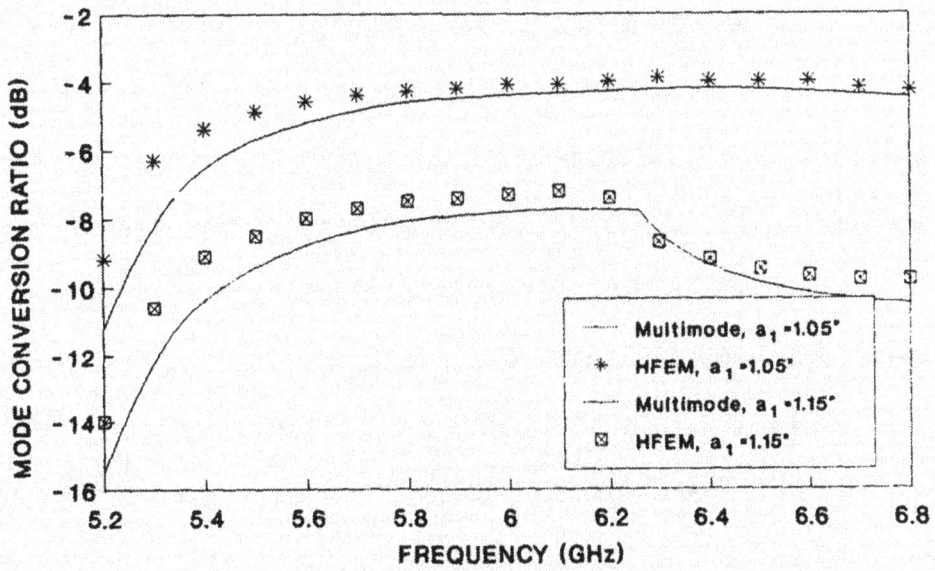

Figure 24. Mode Conversion Ratio for Circular Waveguide Mode Converters: TWOPORT and Multimode [35] Calculations

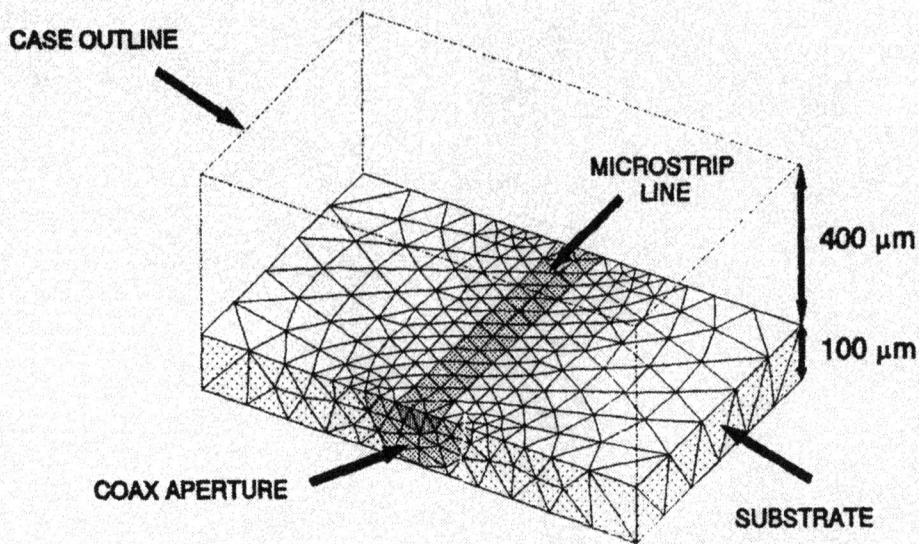

CASE OUTLINE

MICROSTRIP
LINE

400 μm

100 μm

COAX APERTURE

SUBSTRATE

Figure 25. Microstrip Transmission Line Section in Conducting Enclosure

chosen strictly for convenience since the objective of this test was to verify a microstrip transmission line property (guide wavelength).

The substrate dielectric constant was chosen as 12.9 to represent Gallium Arsenide (GaAs). The 75 μm-wide microstrip line has the same characteristic 50 Ω impedance as the coax. The enclosure dimensions were chosen to ensure that there are at least 4 line-widths space between the microstrip and the cavity side and top walls. That is adequate to ensure that the enclosure does not influence the guide wavelength or characteristic impedance [36].

The transmission coefficient for a line length of 500 μm was computed first. Next, the calculation was repeated with finite element geometry scaled by 1.5, giving a line length of 750 μm. The difference in S_{21} phase was 35° at the 40 GHz test frequency, from which the guide wavelength is calculated as 2.57 mm. Quasi-static formulas developed by Wheeler [37] give 2.54 mm. This close agreement indicates that the finite element method is correctly predicting the fields at the interface between highly contrasting dielectrics, even in the presence of a sharp conducting edge.

7.3.2. Microstrip Meander Line. Since the modeling method accurately accounts for all relevant boundary conditions for a simple microstrip line, it should be obvious that it can correctly predict the performance of most passive printed-circuit RF components. A microstrip meander line was chosen as a demonstration case since measured data was available. The circuit dimensions (in μm) and the finite element mesh (substrate cells only) are shown in Figure 26.

The measurement reference planes are at the centers of the pads at each end of the circuit, while those used in the calculations are at the points where the lines begin to taper from 75μm down to 25μm, so the difference must be accounted for when comparing the data. The circuit layout shows the location of via holes that provide a ground for the measurement probes. At the contact points, the 75μm center conductor and 50μm-wide slots form a 50Ω coplanar waveguide (CPW). Using formulas given by Rowe and Lao [38], the effective relative permittivity for that transmission line structure was found as ϵ_{eff}=6.29. The measured S_{21} phase was corrected by adding $360° \cdot (75\mu m)(\epsilon_{eff})^{1/2}/\lambda_0$. The calculated S_{21} phase was also corrected by subtracting the excess phase introduced by the coaxial connectors, computed from a straight length of microstrip line as discussed in the preceding section. Figure 27 compares the measurements and calculations over the frequency range from 1 to 26 GHz. The slight discrepancy in the phase slope is attributable to the differing reactances of CPW-microstrip (measurement) and coax-microstrip (calculations) transitions.

7.4. Importance of Higher Order Modes

An important issue is whether there is in fact any benefit to using higher-order modes in conjunction with a finite element solution. The alternative is to extend the finite element mesh into the connecting waveguides far enough that any higher modes excited by the cavity's contents and/or apertures have decayed to insignificance [39]. The answer depends on two factors: first, the mode excitation strengths; and second, their attenuation constants in the waveguides, which

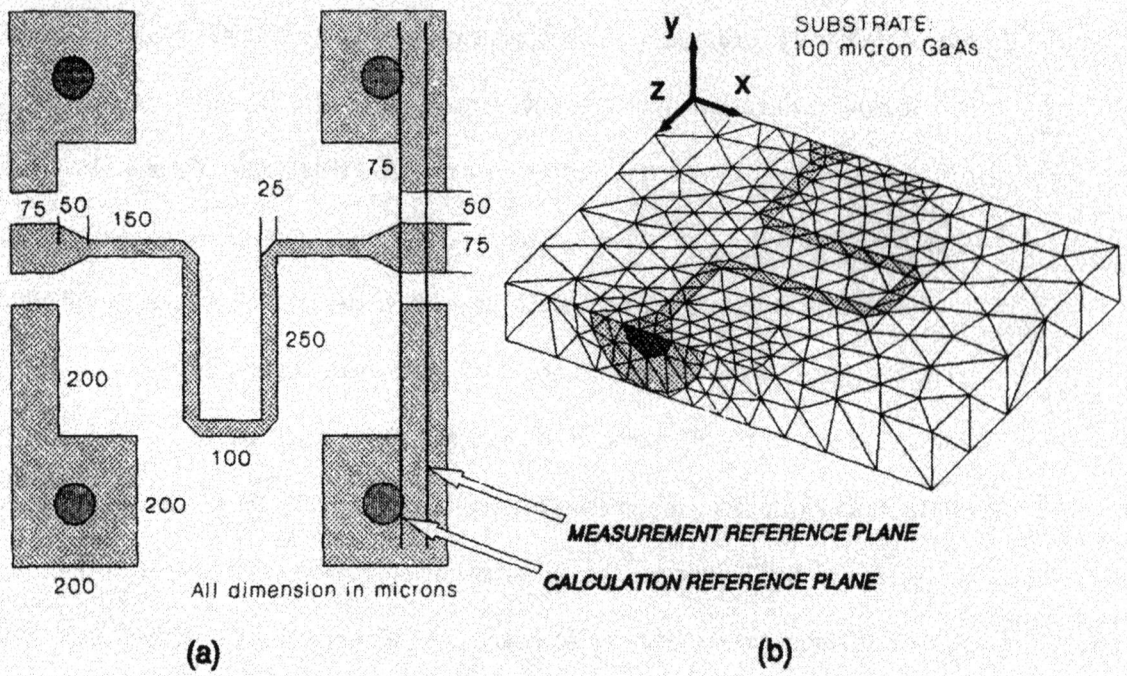

Figure 26. Microstrip Meander Line: (a) Wafer Metallization Dimensions (in μm); and (b) Finite Element Mesh for Substrate

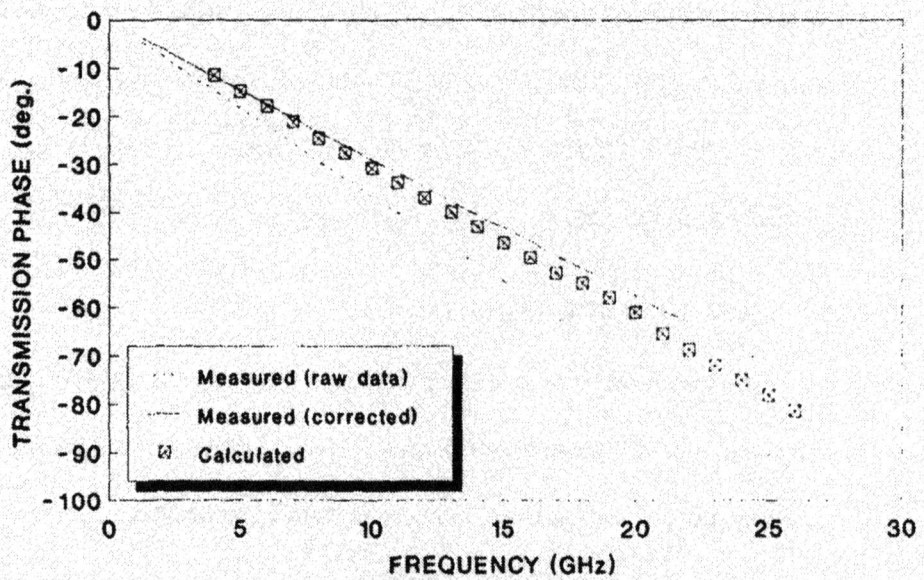

Figure 27. Comparison of Measured and Calculated S_{21} Phase for Microstrip Meander Line

depends mainly on the ratio of the frequency to the mode cutoff frequency.

Figure 28 shows how the magnitude of S_{11} converges with the number of modes in the one-quadrant coaxial iris problem (Figures 21 and 22). The most important modes are the TEM and the TE_{11} modes. The latter is excited at about -10 dB, and its cutoff frequency in APC-7 coax is about 20 GHz. At 18 GHz it decays by 29 dB per wavelength, so in order for it to decay below -30 dB (regarded as negligible for purposes of S parameter calculation), the mesh would need to extend roughly $\frac{2}{3}\lambda$ into the waveguide in each direction. The finite element mesh and the interaction matrix would then contain an additional 500-1500 edges and unknowns, respectively.

On the other hand, the two microstrip problems do not excite any higher order modes above the -30 dB level, so those devices could be accurately modeled using only the lowest-order mode and still terminating the mesh at the waveguide aperture. Thus, the use of the mode sum as a continuity condition can make the finite element solution more efficient, but the improvement

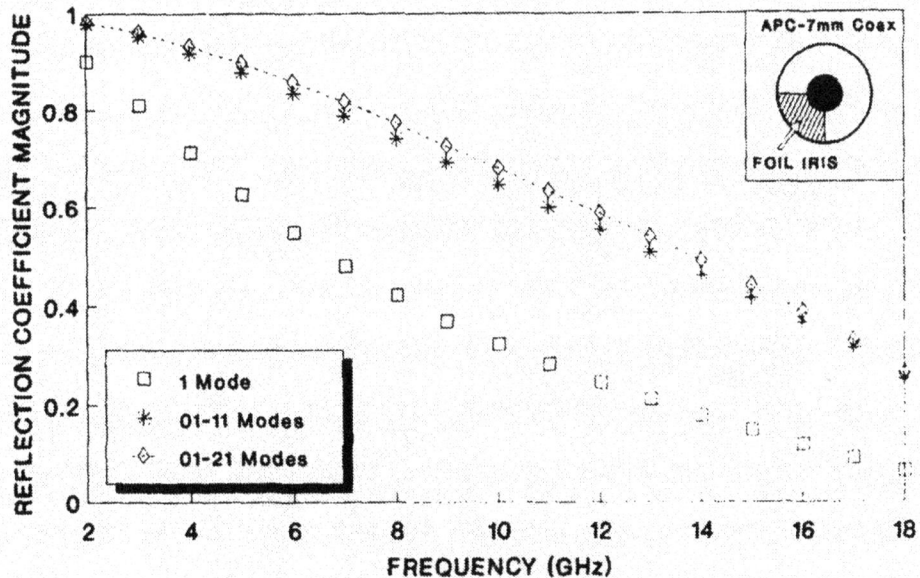

Figure 28. Reflection Coefficient for One-Quadrant Coaxial Waveguide Iris
Calculated with Varying Numbers of Higher-Order Modes

over the dominant-mode-only approach is problem-dependent.

7.5. Summary

In addition to validating the essential properties of the electromagnetic solution approach, the TWOPORT code provided valuable data on *practicality*: typical mesh sizes and execution times. The data, summarized in Table I, includes all of the test cases discussed in this chapter, plus three others: The rectangular and circular waveguide cases were short sections of waveguide used to verify dominant mode propagation; and the coax-to-rectangular waveguide transition consists of a right-angle metal launcher in a section of rectangular waveguide approximately $\lambda/3$ long. The meshes for these three problems are shown in Chapter VIII, where they are used as cavity array elements.

Table I. Mesh Sizes, Iterations and Matrix Solve Time for TWOPORT Test Cases
(Solve time is CPU time on VAX® 8650 minicomputer)

Test Case	Volume/λ^3	Number Mesh Cells	Cells/λ^3	Number of Iterations	Solve Time (min.)
Rectangular Waveguide	.149	348	2067	.9N	0.7
Circular Waveguide	.120	342	2850	1.3N	1.4
Iris in Coaxial Waveguide	.0135	847	63×10^3	.9N	12.5
Circular Waveguide Mode Converter	.376	1504	4000	0.5N	43.5
Coax-to-Rectangular Waveguide Transition	.294	2442	8294	4.7N	226.
Microstrip Meander Line	9.4×10^{-5}	4190	4.5×10^7	8.4N	524.
Microstrip Transmission Line	5.7×10^{-4}	4662	8.2×10^6	7.1N	683.

The first column in Table I gives the total volume in cubic wavelengths for each problem. For the simpler problems, the number of cells is determined primarily by the sampling requirement of 10 nodes per wavelength. More complicated structures, however, require more cells in order to capture fine details of geometry. Column 4 gives the number of iterations in terms of N, the number of mesh edges, that were required to obtain convergence of the residual norm to .001 of its initial value (using the zero vector as an inital guess). In all cases, the maximum number of iterations was less than 10 times the number of edges (unknowns). The matrix solve time per frequency sample is given in the last column. These results indicate that practical design problems can be accomplished on typical minicomputers and workstations.

This chapter has validated two key elements of the solution approach. First, it showed that FEM can correctly account for boundary conditions typical of antenna problems: conductors, conductor edges and dielectric interfaces. Second, it showed the validity and usefulness of the waveguide mode integral equation as a continuity condition. These results validated not only the generic concepts, but the computer code implementation as well. The latter was especially important since the TWOPORT code included most of the structure and modules needed for the subsequent cavity array problem solution discussed in the next chapter.

VIII. Validation - Cavity Array Problem

The second development stage replaced the integral equation for the outlet waveguide with the periodic integral equation. This allows validation of the periodic radiation condition without the complexity of side-wall periodicity conditions. It also provides a tool useful for analyzing the properties of radiators such as open-ended waveguides, cavity-backed slots and multimode horns. This chapter describes its implementation in a computer code and presents comparisons of its calculations with published results. Also presented are two specific examples of radiators that are beyond the capability of previous methods: a pyramidal horn; and a rectangular waveguide with a coaxial transition in close proximity to the aperture.

8.1. Computer Code Implementation

A FORTRAN program named CAVIARR (cavity array) implements the solution to this generic problem, depicted in Figure 7. Figure 29 is an outline of its actions. Note that all of these actions up to step IV.C. are essentially the same as in program TWOPORT. The interior finite element matrix and the inlet waveguide submatrix calculations are exactly the same, since they do not depend on the scan angle. For each separate scan angle, the submatrix due to the radiating aperture must be recalculated, then the system must be solved again for the field vector E. The active reflection coefficient calculation is the same as the calculation for S_{11} in TWO-PORT. However, instead of the S_{21} calculation, CAVIARR calculates the element far field and transmission coefficients into θ and ϕ polarizations.

8.2. Waveguide Arrays

8.2.1. Rectangular Array. The scan-dependent reflection coefficients for several open-ended waveguide arrays are available from publications by other authors. For example, the properties of the rectangular waveguide array shown in Figure 30a were first presented by Dia-

I. READ INSTRUCTIONS AND OPTIONS

II. READ PROBLEM GEOMETRY

III. CREATE EDGE-BASED GEOMETRY

IV. FOR EACH FREQUENCY:

    A. COMPUTE TERMS OF S^{EE} ACCORDING TO (20), (22)

    B. FOR WAVEGUIDE A:

        1. COMPUTE INCIDENT CURRENT VECTOR IN (32)

        2. COMPUTE TERMS OF S^{EJ} FROM (34), (35)

    C. FOR EACH ANGLE:

        1. COMPUTE TERMS OF S^{EJ} FROM (51)

        2. SOLVE $(S^{EE} + S^{EJ})\, E = E^{inc}$ FOR E

        3. COMPUTE REFLECTION COEFFICIENT AND MODE
           EXCITATION COEFFICIENTS FROM (36)

        4. COMPUTE ELEMENT FAR FIELD AND TRANSMISSION
           COEFFICIENTS FROM (52)-(55)

Figure 29. Solution Procedure in Program CAVIARR

mond [40] and independently confirmed by several others [29],[41].

Figure 30b is the finite element model used by program CAVIARR - simply a short section of waveguide. The nodes on the shaded surfaces were tagged as conductors, while those on the front and back faces were tagged as radiating aperture and waveguide boundaries, respectively.

Figure 31 compares Diamond's results with CAVIARR computations. The interesting feature of this test case is the scan blindness near 33° in the H-plane (the $\phi=0$ plane). When the array scans to that angle, nearly all of the incident power is reflected back into the waveguide.

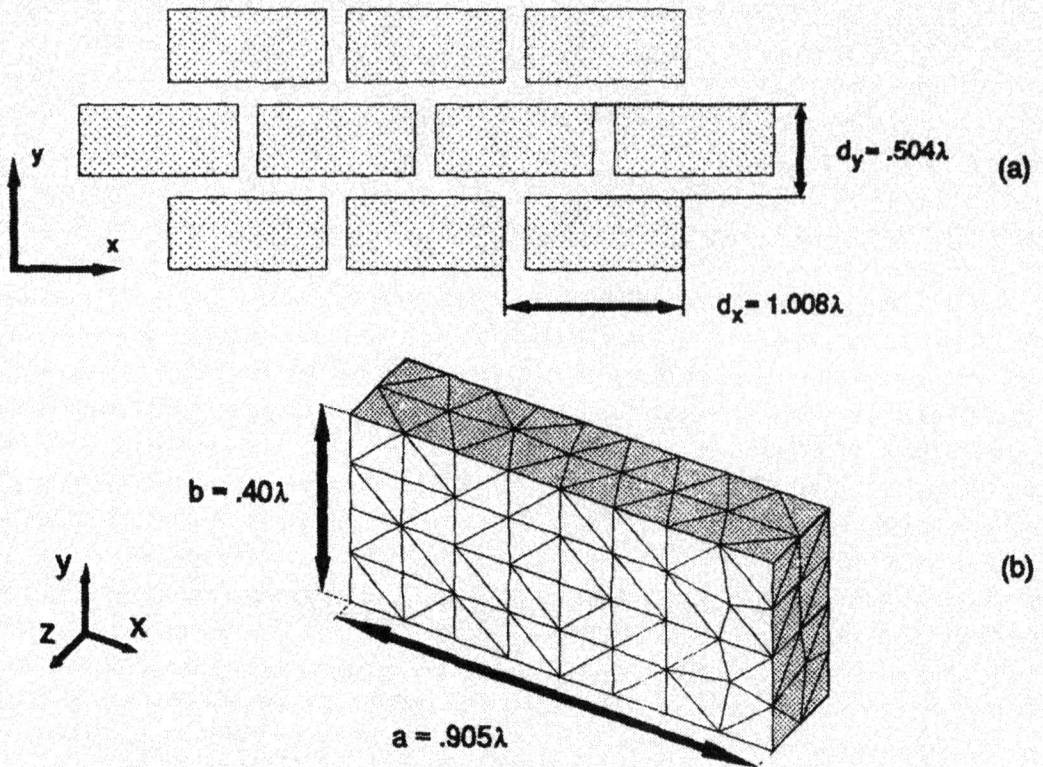

Figure 30. Rectangular Waveguide Test Case: (a) Array Lattice; and (b) Tetrahedron Mesh

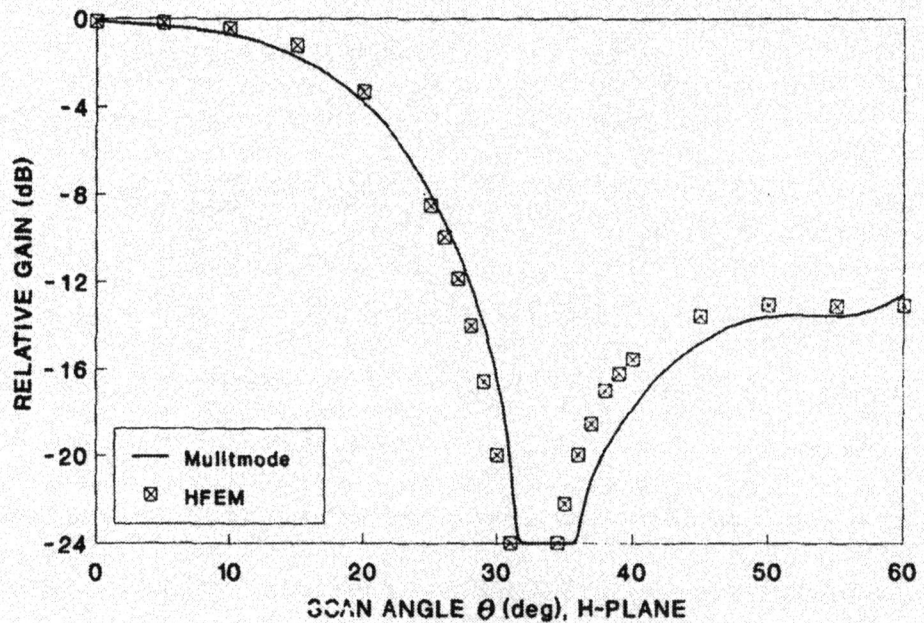

Figure 31. H-Plane ($\phi=0$) Active Element Gain vs. Scan Angle for Rectangular Waveguide Test Case

The CAVIARR (HFEM) calculations used waveguide modes up to $m,n=2,3$ and Floquet modes up to $|m|,|n|=5,5$. This number is consistent with the estimate given earlier in Section 5.3 since the mesh spacing in the aperture is approximately $\lambda_0/10$. In fact, $|m|,|n| \le 4$ was adequate to ensure convergence of the active reflection coefficient to within .1% in magnitude and 1° in phase.

8.2.2. Rectangular Array with Conducting Iris. A straightforward method for eliminating the H-plane scan blindness in the rectangular waveguide array is to place conducting irises in the apertures. For example, Lee & Jones [41] performed a multimode[1] analysis for the same array lattice and waveguide size as in Figure 30, except that a thin conductor blocked the left and right ¼ of each aperture. The same finite element mesh as before (Figure 30b) was used for this problem, except that the nodes associated with the iris were tagged as separate conductors. Figure 32 compares the CAVIARR calculations with the H-plane element gain pattern from [41]. The scan blindness was indeed suppressed by the addition of the irises, but at the expense of a reduction in the broadside gain.

The close agreement between the CAVIARR calculations and published results for these two test cases demonstrates that the hybrid periodic integral equation/finite element formulation is valid. It also demonstrates that its implementation in the computer code is accurate.

8.2.3. Circular Waveguide Arrays. The lattice geometry for an array of circular waveguide radiators is shown in Figure 33a. The incident waveguide field is in the dominant TE_{11} mode and polarized parallel to the y axis. Multimode calculations for this geometry are available from Amitay et. al. [29].

---

[1] The *multimode* method is sometimes referred to as a moment method. It equates the transverse fields on the two sides of the aperture as sums over modes appropriate to each region. The modes of one waveguide are used as testing functions in order to generate a matrix equation. The generic technique developed for waveguide discontinuities was adapted to the phased array problem using Floquet modes.

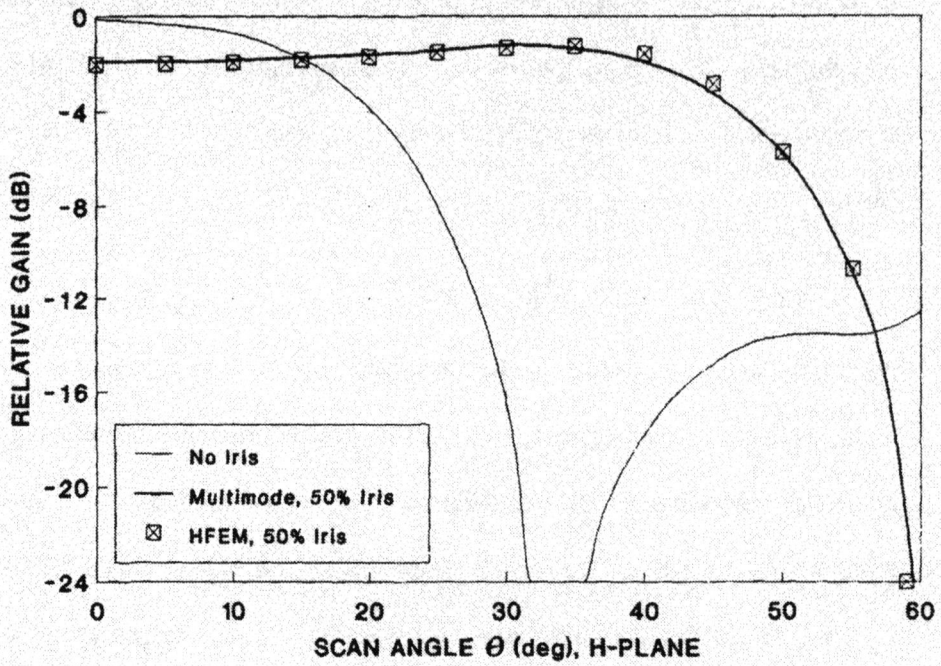

Figure 32. Active Reflection Coefficient vs. Scan Angle for Rectangular Waveguide
Phased Array with Conducting Irises

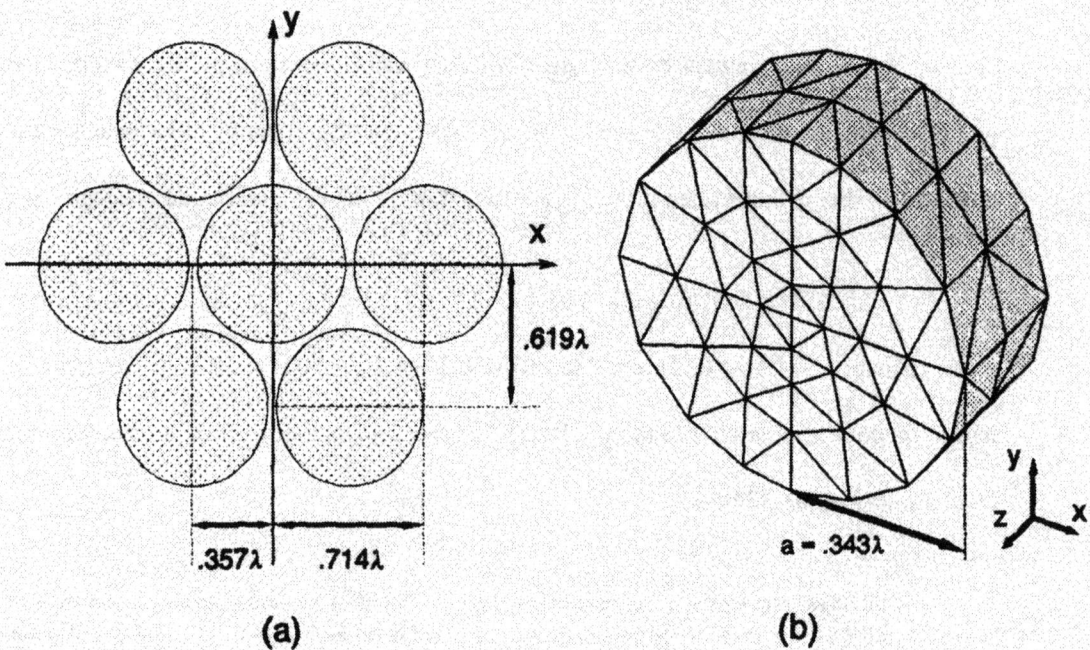

Figure 33. Circular Waveguide Array Test Case: (a) Lattice; and (b) Tetrahedron Mesh

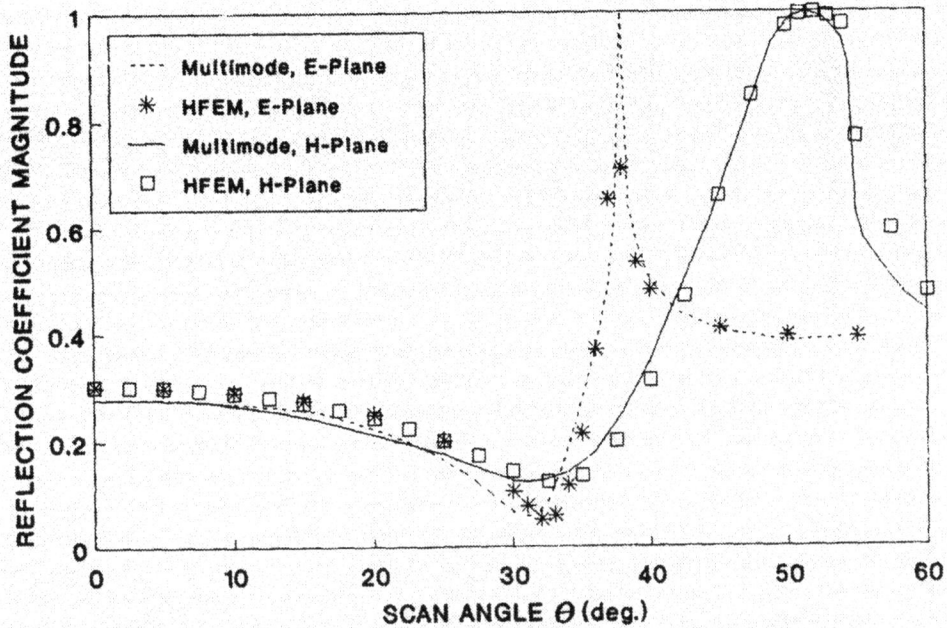

Figure 34. Circular Waveguide Array Active Reflection Coefficient vs. Scan Angle

The finite element mesh used for CAVIARR is shown in Figure 33b. Its radius was adjusted so that the sum of the tetrahedron volumes was the same as a perfect circular cylinder with $.343\lambda_0$ radius. Figure 34 shows the finite element calculations along with the published multimode results for both E- and H-plane scanning [29:276,280]. Again, the CAVIARR code is accurately predicting the active reflection coefficient at all scan angles.

A method for suppressing the scan blindness in circular waveguide arrays is to use dielectric-loaded waveguides that can be packed closer together since their radius is smaller for a given wavelength. In the following test case, the waveguide radius was adjusted so that $\epsilon_r^{1/2} a = .343\lambda_0$. The lattice spacing was adjusted so that d_x/a and d_y/a are constant. The finite element model was simply a scaled version of the mesh in Figure 33b. Figure 35 is a comparison of the loaded and unloaded cases, the latter with $\epsilon_r=4.1$, for H-plane scanning [29:290].

A third and final test case involving circular waveguide arrays also includes a dielectric,

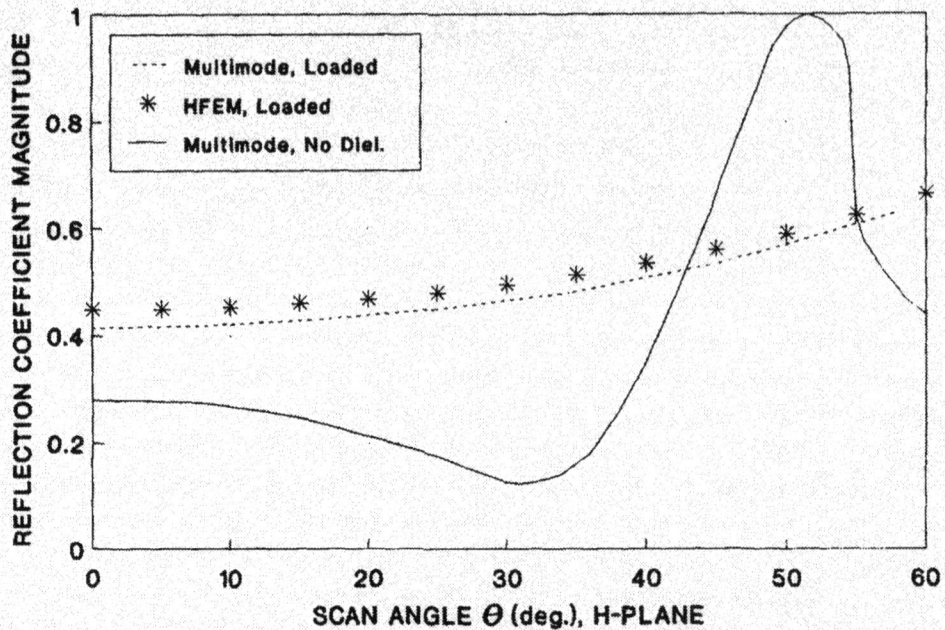

Figure 35. Active Reflection Coefficient vs. Scan Angle - Circular Waveguide
Array with and without Dielectric Loading

but instead of completely filling the waveguide, it is only a short plug in the aperture end, flush

with the ground plane. Amitay et. al. give results for E-plane scanning using the lattice in Figure

33a, and $.429\lambda_0$ length plugs with $\epsilon_r = 1.8$ [29:293]. The finite element model for this problem

was a longer version of that depicted in Figure 33b - it was 6 mesh cells in length, in order to

meet the $\lambda/10$ sampling requirement. Figure 36 compares those results for the no-dielectric case.

The results of the CAVIARR and multimode computations are substantially identical. (There are

sources of error in both approaches, such as the number of modes and the number of integration

points. The important fact is that both methods are accurate enough to allow a judgement of

whether or not an antenna design is acceptable.) This demonstrates the important capability of

the hybrid finite element method for modeling arrays that have dielectric loading. Unlike the

multimode method, it is not restricted to homogeneous dielectrics.

66

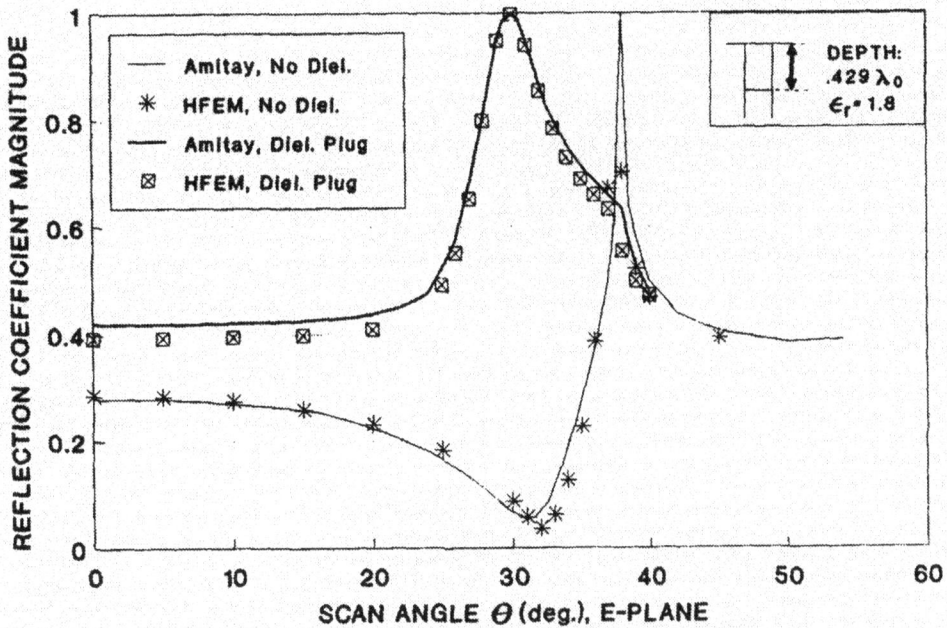

Figure 36. Active Reflection Coefficient vs. Scan Angle - Circular Waveguide
Array with and without Dielectric Plugs

8.3. Pyramidal Horn Array

The multimode method was the only rigorous technique available for addressing cavity array problems. This confined the available solutions to structures for which mode sets can be constructed, mostly cylindrical waveguides. There is, however, one example by Amitay & Gans [42] that uses a variation of the multimode method to approximate the scanning characteristics of an array of pyramidal horns. Their technique models the pyramidal horn, shown in Figure 37a, as a rectangular waveguide whose dimensions are the same as the horn mouth, containing planes at various z locations that are transparent to those waveguide modes that may propagate, and shorting those that are cut off. In spite of the approximate nature of their prediction technique, they obtained fairly good agreement with measured data.

This problem is a fairly stressing case for the hybrid finite element method. First of all, the mesh, depicted in Figure 37b, has approximately 6000 edges. Furthermore, the large unit

67

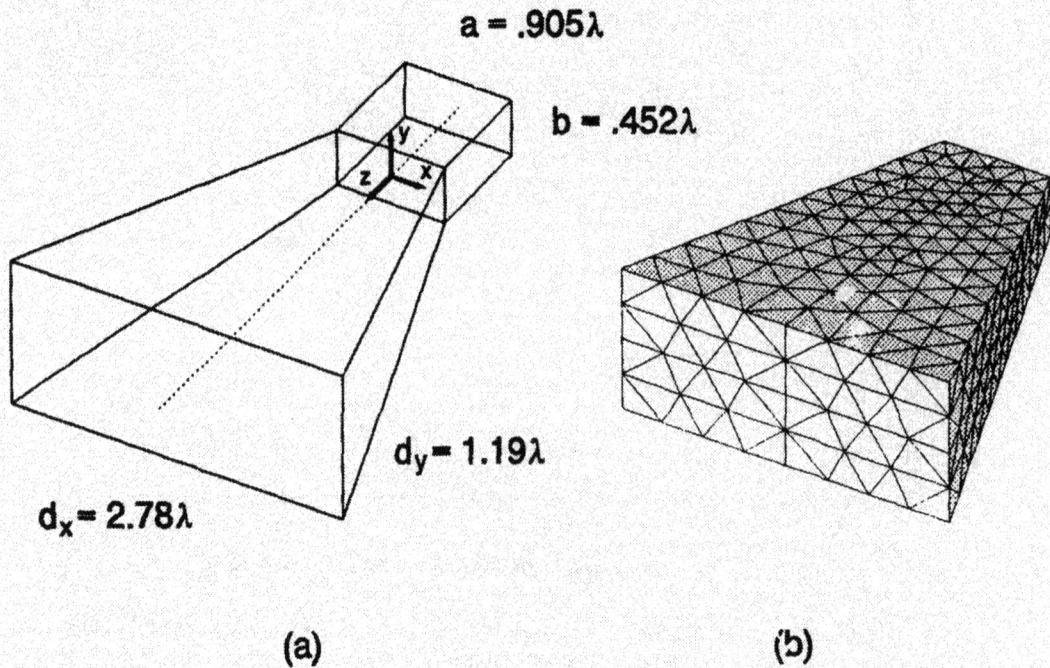

Figure 37. Pyramidal Horn Radiator: (a) Dimensions; (b) Tetrahedron Mesh

cell size (the same as the horn mouth) requires an unusually large number of Floquet modes (60 in k_x and 15 in k_y).

Figure 38 is a comparison of the CAVIARR calculations for active element gain with measured data from [42]. The scan plane is $\phi=90°$, for which the co-polarized field is the E_θ component. The gain in decibels is $20 \log_{10} E_\theta$. The interesting feature of this test case is the scan blindness near 40°, due to excitation of a higher-order waveguide mode at the aperture. In contrast to the waveguide cases discussed earlier, the active reflection coefficient does not become large near the blindness angle. Since there is more than one propagating Floquet mode in this instance, incident power that is not transmitted in the desired direction is transmitted into another mode, i.e. a grating lobe.

8.4. Coaxial-to-Rectangular Waveguide Launcher

An end-wall transition, or "launcher," from coaxial to rectangular waveguide is shown in Figure 39. This type of transition has an advantage in a rectangular waveguide phased array because the waveguides are packed too closely to use a broad-wall launcher. This design, due to Tang and Wong [43], is an extension of the coax center conductor with a shorting post joining it to the broad wall. The probe is centered in height and offset 1/10th the waveguide width. Its length should be approximately one fourth the guide wavelength at the design frequency. For a length of 9mm, the frequency response is shown in Figure 40 (computed by program TWO-PORT).

The experimental array reported in [43] had a long waveguide section between the launchers and the open apertures, and higher-order waveguide modes excited by either one would not affect the other. The antenna's weight and size would both benefit if the launchers were placed as close as possible to the apertures, but then their mutual interactions are not negligible.

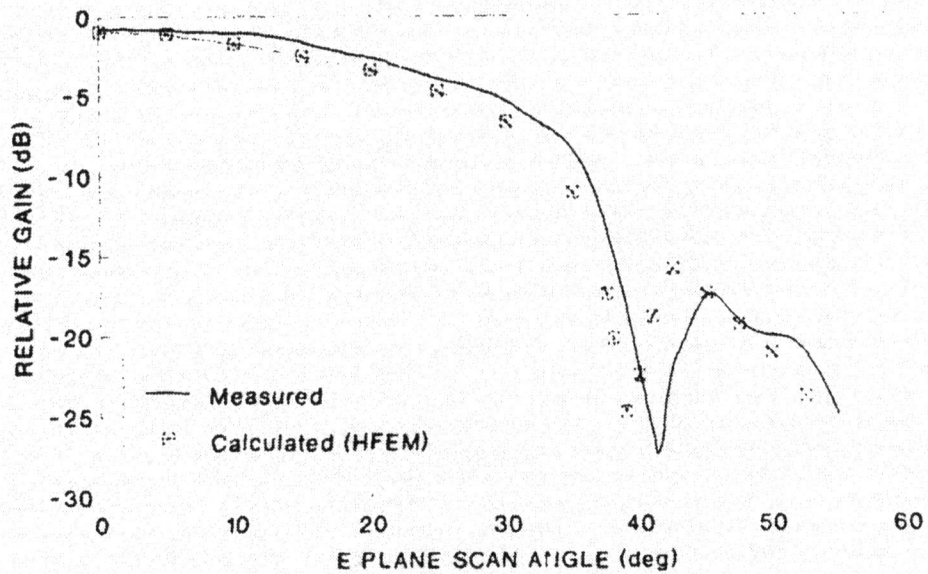

Figure 38. Active Element Gain vs. E-Plane Scan Angle for Pyramidal Horn Array (Measured Data from Amitay & Gans [42])

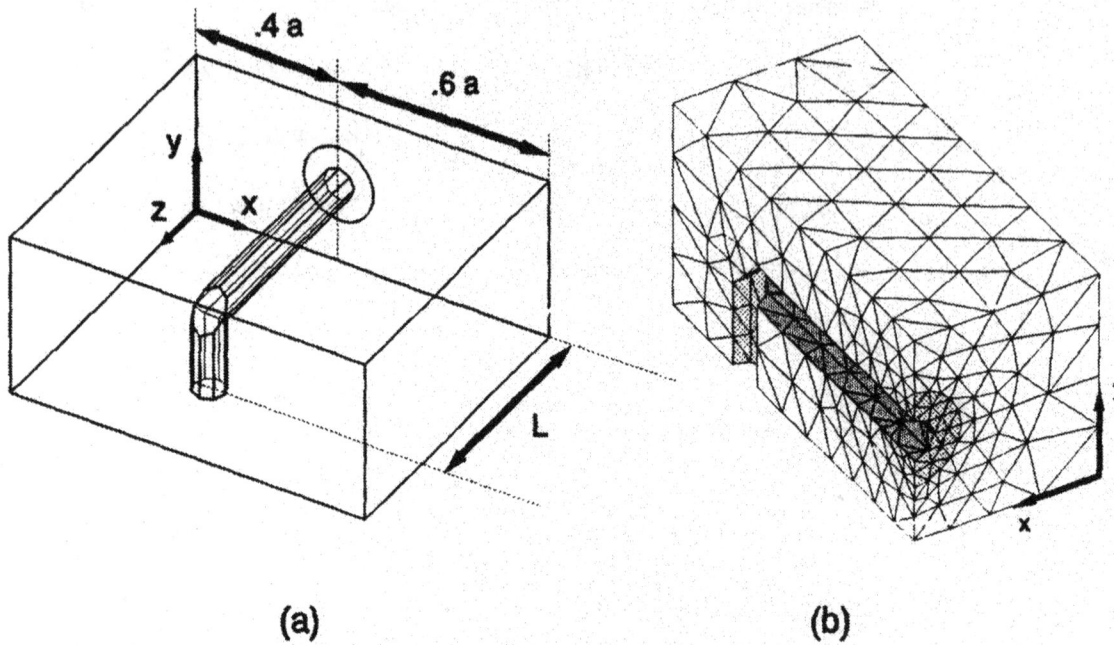

Figure 39. End-Wall Transition from Coax to Rectangular Waveguide (After Tang & Wong [43]): (a) Geometry; (b) Cutaway of Tetrahedron Mesh

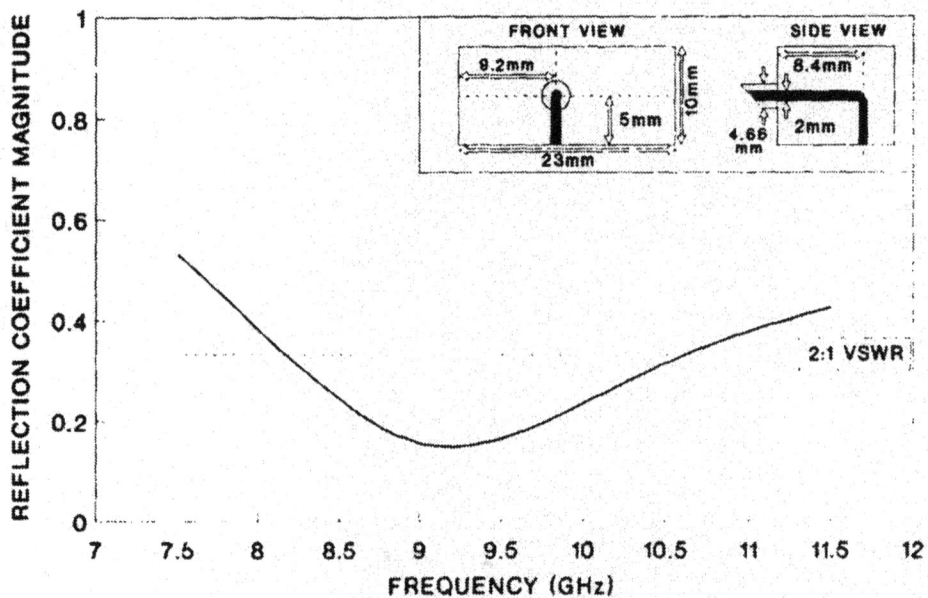

Figure 40. Reflection Coefficient (S_{11}) for Coax-Rectangular Launcher: Probe Length = 9mm; Post Height = 5mm; X-Band Waveguide

Considering again the rectangular waveguide test case discussed in Section 8.2.1, suppose the waveguide elements are only 10mm deep, with a launcher extending to within 2.5mm of the aperture. The predicted active reflection coefficient vs. angle is shown in Figure 41, compared to the original case, where the array is fed by semi-infinite rectangular waveguides (both calculations by CAVIARR). Evidently, the launcher in close proximity to the waveguide opening prevents the formation of the higher-order waveguide mode that is responsible for the scan blindness condition, which was the purpose of the conducting irises discussed by Lee & Jones. Furthermore, the interactions between the launcher and the waveguide aperture do not generate any additional resonance effects, so there is no need to separate the two by a long length of waveguide. Thus, the array element consisting of a coaxial launcher in a short waveguide section is a better solution than was previously available because it achieves the same result with a smaller and simpler structure. Although rectangular waveguide arrays are outdated, this is nonetheless an illustration that: (a) the hybrid finite element method can be used to solve practical

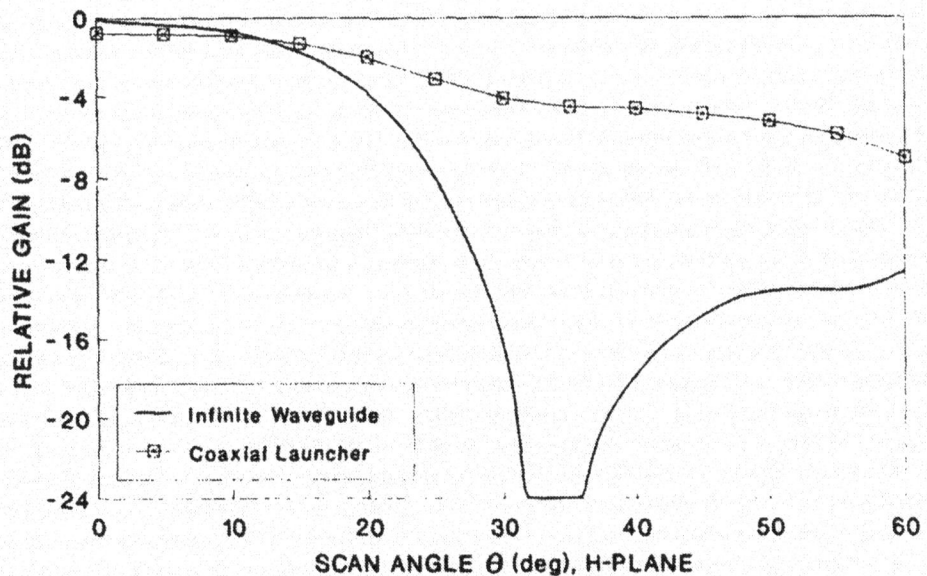

Figure 41. Active Reflection Coefficient for Rectangular Waveguide Array with Coaxial Launcher and Rectangular Waveguide Feeds

problems for which accurate methods were not available before; and (b) the cavity array solution has important applications by itself even though it, like TWOPORT, was an intermediate step in the code development. The following chapter describes the implementation and results from the final step, which implements the periodicity conditions at unit cell side walls for general array radiators.

IX. Validation - General Array Problem

The third and final stage of computer code development and validation implemented the periodicity condition for unit cell side walls, removing the requirement that the radiators be separated by conducting walls. Initial validations were performed for open waveguide arrays using the previous chapter's results. Test cases for arrays of microstrip patches and monopole arrays were used for further validation, the latter including a hardware experiment. Last, performance predictions for flared notch and printed dipole arrays are presented to demonstrate the method's versatility.

9.1. Computer Code Implementation

A FORTRAN program named PARANA (phased array antenna analysis) implements the general array problem solution. Figure 42 is an outline of the program's actions. Up to step IV.C. it is essentially identical to CAVIARR, except that a listing of image edges is created in step III. These identify the -x and -y boundary counterparts for each edge on the +x and +y unit cell walls. This places an additional requirement on the geometry file - the unit cell boundary nodes must be tagged so that these edges can be identified. In addition, it requires the mesh generator to create the grid in such a way that the surface mesh on opposing faces is identical. Although the CAD software (I-DEAS^{TM} [44]) does not have any provision for enforcing this requirement, it results naturally when the mesh areas and constituent curves are defined in consistent and logical order.

In step IV.C.1., the calculation for the radiation boundary terms of S^{EJ} is slightly modified to include special handling for -x and -y boundary edges, as discussed in Section 6.4. Step IV.C.2. is a straightforward implementation of the algorithm shown in Figure 17. One of its effects is to zero out the rows and columns of S^{EE} corresponding to +x and +y boundary edges.

73

I. READ INSTRUCTIONS AND OPTIONS

II. READ PROBLEM GEOMETRY

III. CREATE EDGE-BASED GEOMETRY AND LIST OF IMAGE EDGES

IV. FOR EACH FREQUENCY:

    A. COMPUTE TERMS OF S^{EE} ACCORDING TO (20), (22)

    B. FOR WAVEGUIDE A:

        1. COMPUTE INCIDENT CURRENT VECTOR FROM (33)

        2. COMPUTE TERMS OF S^{EJ} FROM (34)

    C. FOR EACH ANGLE:

        1. COMPUTE TERMS OF S^{EJ} FROM (51) USING OVERLAP ELEMENTS

        2. IMPOSE PERIODICITY CONDITION ON S^{EE} ACCORDING TO FIG. 17

        3. ELIMINATE ZERO ROWS & COLUMNS FROM $(S^{EE} + S^{EJ})$

        4. SOLVE $(S^{EE} + S^{EJ}) E = E^{inc}$ FOR E

        5. RESTORE BOUNDARY EDGES

        6. COMPUTE REFLECTION COEFFICIENT AND MODE EXCITATION COEFFICIENTS FROM (36)

        7. COMPUTE ELEMENT FAR FIELD AND TRANSMISSION COEFFICIENTS FROM (52)-(55)

Figure 42. Solution Procedure in Program PARANA

Step IV.C.1. had a similar effect on S^{EJ}. Before starting the matrix solution, the matrix is compressed to eliminate zero rows and columns. After the matrix is solved, the periodicity conditions are used to solve for the field along the $+x$ and $+y$ boundary edges.

9.2. Waveguide Arrays

The PARANA code is also capable of modeling open-ended waveguide arrays, although less efficiently than CAVIARR. A comparison between the two was a means for verifying the periodic boundary condition while keeping all other details of the problem and solution the same. Figure 43 is the tetrahedron mesh used as the input to PARANA. It is a shallow slice of the unit cell in free space *outside* the waveguide, above the ground plane. The shaded area identifies those nodes and cells bordering on the ground plane. Contrast this to the CAVIARR geometry model (Figure 33b), which is a slice of waveguide below the ground plane.

The test case parameters were: $a = .3\lambda_0$, $d_x = .88\lambda_0$; $d_y = .733\lambda_0$ and $\gamma = 60°$ (equilateral lattice). Figure 44 compares the computed results from the two codes for scanning in the H-plane (x-z plane). The fact that they are essentially identical is evidence that the periodic boundary condition is working as proposed. The results for E-plane scanning were similar, with the same degree of agreement between the two codes.

A second waveguide test case was a rectangular lattice of rectangular waveguides. The lattice dimensions were the same as the waveguide size, i.e.: $a = d_x = 23mm$ and $b = d_y = 10mm$. The finite element mesh was similar to Figure 30 except that there were three layers of cells, each with the same thickness. The perimeter edges of the first layer were tagged as con-

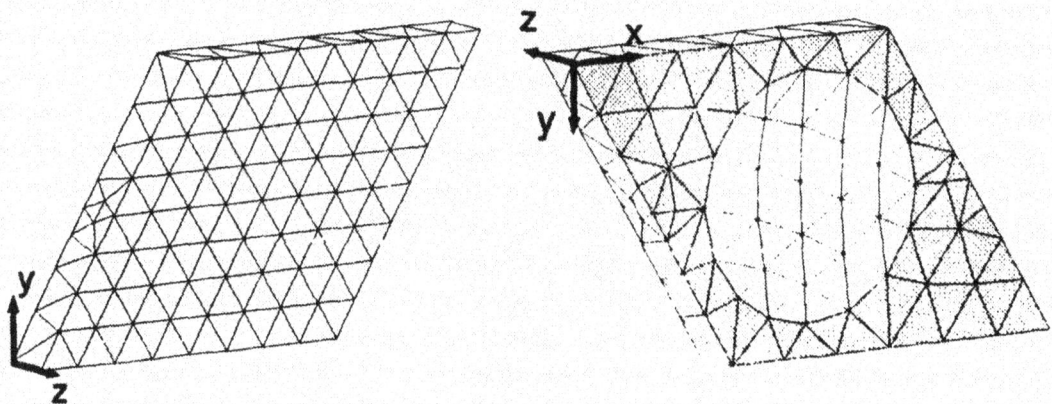

Figure 43. Finite Element Mesh for a Skewed-Lattice, Circular Waveguide Array Unit Cell

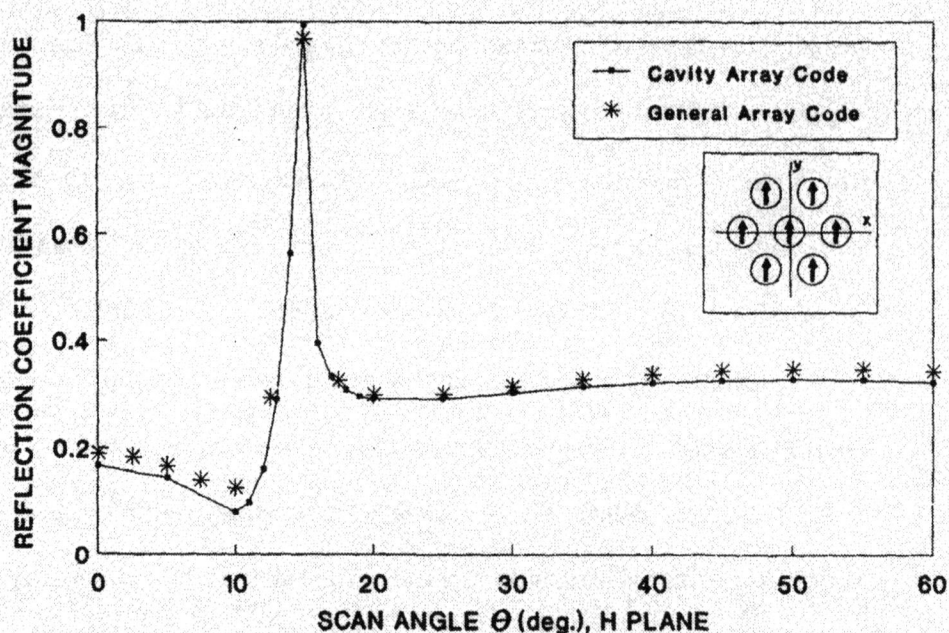

Figure 44. Circular Waveguide Array Active Reflection Coefficient - Comparison of Results Using Cavity Array (CAVIARR) and General Array (PARANA) Models

ductors to indicate that the mesh extends part way into the waveguide. The perimeter edges of the other two layers were tagged as unit cell walls. The calculations by PARANA for 10 GHz, shown in Figure 45, are again essentially the same as the CAVIARR results. This test case illustrates some potentially important flexibilities of the implementation: the mesh may extend into the feed waveguide; and the feed waveguide may adjoin the unit cell boundary.

9.3. Microstrip Patch Array

The first of four demonstration cases uses a simple rectangular microstrip patch on a substrate that is thick enough to support a surface wave in the dielectric. The geometry, shown in Figure 46, has the dimensions (in wavelengths) used by Pozar in MoM calculations [45]. The substrate has relative permittivity of 12.8 to represent GaAs. Since the input impedance of the patch is very low (less than 5Ω) due to the substrate thickness, it presents a large mismatch to the coaxial feed. Consistent with [45], the active reflection coefficient is normalized using:

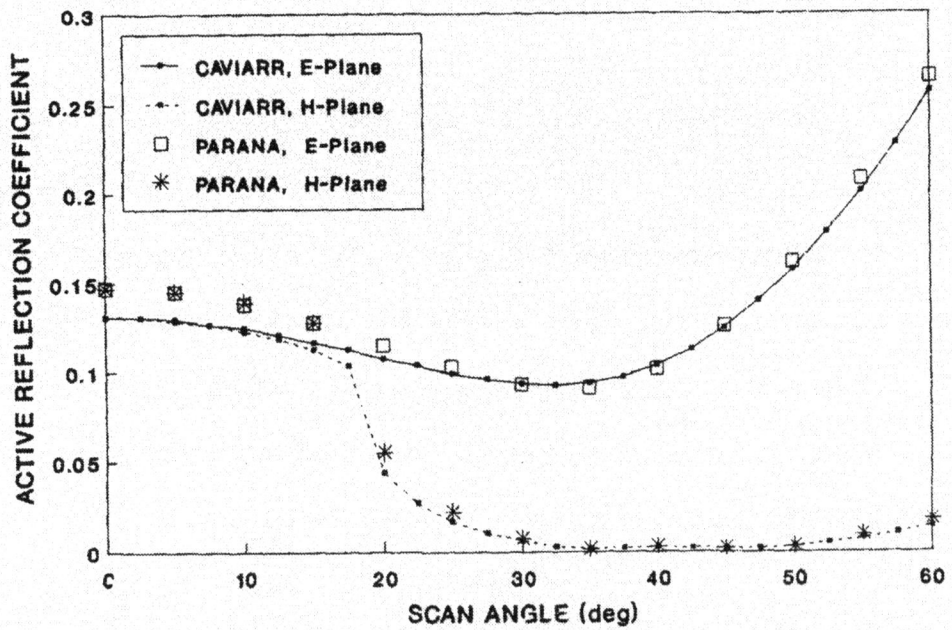

Figure 45. Rectangular Waveguide Array Active Reflection - Comparison of Results
Using Cavity Array (CAVIARR) and General Array (PARANA) Models

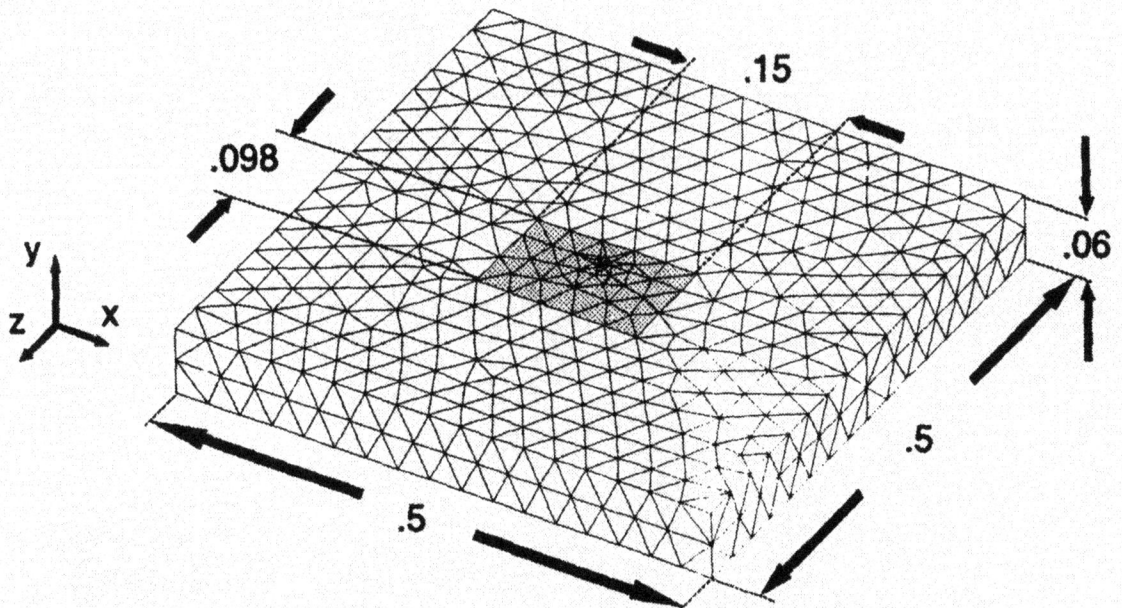

Figure 46. Microstrip Patch Radiator (dimensions in wavelengths)

77

$$R_a(\theta,\phi) = \frac{Z_{in}(\theta,\phi) - Z_{in}(0,0)}{Z_{in}(\theta,\phi) + Z_{in}^*(0,0)} \qquad (65)$$

This measure discounts the effect of the feed impedance mismatch to more clearly reveal the effects of scanning. Figure 47 compares the PARANA (HFEM) and moment method calculations. The agreement is within approximately 7%, with the greatest difference at the angle where there is a scan blindness due to a surface wave in the dielectric slab. This discrepancy may be due to the feed modeling ([45] used an idealized probe feed). The fact that the PARANA code is predicting the existence of the trapped surface wave is a further confirmation that the periodic boundary conditions are effective and correctly implemented.

9.4. Clad Monopole Array Experiment

A candidate antenna design for a space-based radar, proposed by Fenn, was a planar array of monopoles above a ground plane [46]. Each monopole is simply an extension of a

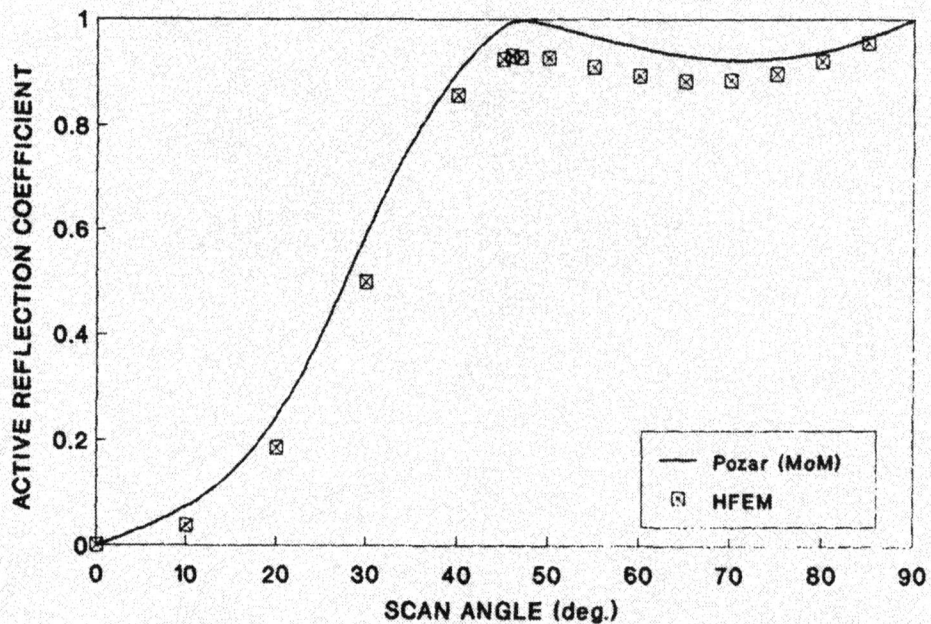

Figure 47. Active Reflection Coefficient for Microstrip Patch Array, E-Plane Scan

coaxial waveguide's center conductor. Fenn's MoM analysis was adequate for the bare mono-pole, but not for one with a dielectric sheath, or cladding. The sheath, which can be simply an extension of the coax insulator, is an important enhancement to the monopole design because it increases the bandwidth substantially.

9.4.1. Initial Validation. For a first validation check, PARANA calculations were compared to Fenn's calculations for $.25\lambda_0$-long monopoles arrayed in a $.5\lambda_0$ x $.5\lambda_0$ lattice. Figure 48 shows the active reflection coefficient vs. scan angle. The essential scanning charac-teristics are verified, although there is some discrepancy at the angles where the reflection coef-ficient is very small. This is at least partly due to a simplification in Fenn's MoM model: It used an assumed form of current distribution (piecewise sinusoidal) on the monopole that is not allowed to change with scan angle. Herper and Hessel used the same simplification, and showed a large disparity between calculated and measured results in the vicinity of 60° [47].

9.4.2. Bandwidth Enhancement with Cladding. Figure 49 shows one half of the tetrahe-dron mesh used by PARANA for the following test case: The monopole (15.875mm long) is represented as a void extending through the mesh from top to bottom. Shading has been added to the figure to identify the cells comprising the dielectric cladding (Teflon, $\epsilon_r = 2.1$). The remaining mesh cells are free space. The lattice is triangular with $d_x = 36$mm and $d_y = 31$mm (nearly equilateral). This interelement spacing allows scanning over the $\theta \leq 90°$ hemisphere without grating lobes for frequencies up to 4.8 GHz.

Figure 50 shows contour plots of $|R_a|$ vs. scan angle and frequency for clad and unclad monopoles. These calculations are for the $\phi = 90°$ scan plane, but similar results would be ob-served for any scan plane due to the equilateral lattice. Both arrays achieve an acceptable reflec-tion coefficient ($|R_a| \leq .33$, or VSWR $\leq 2:1$) over only a very limited range of scan angles near

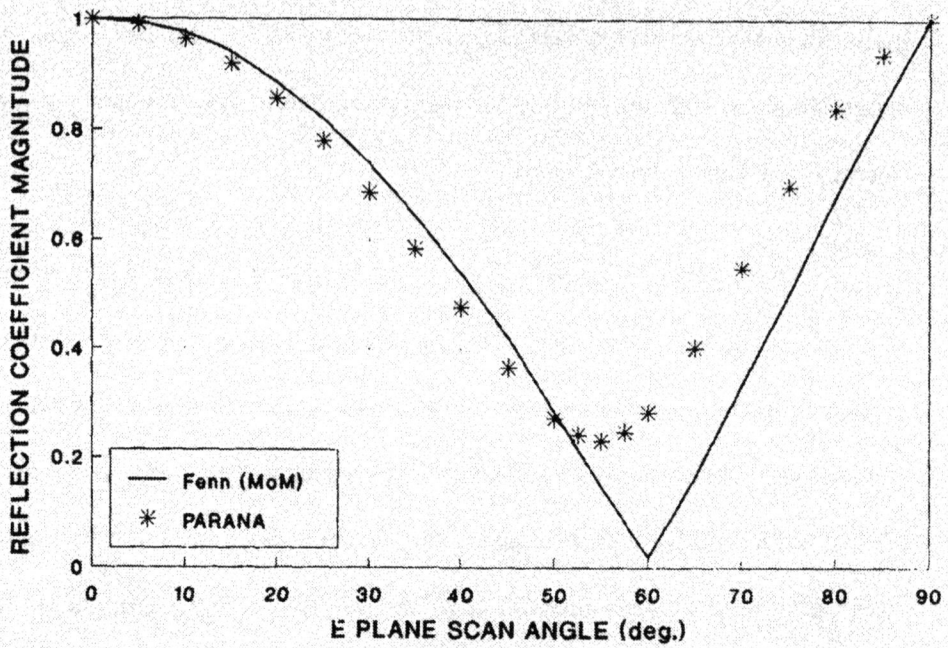

Figure 48. Reflection Coefficient vs. Scan Angle ($\phi=0$ plane) for a Monopole Array

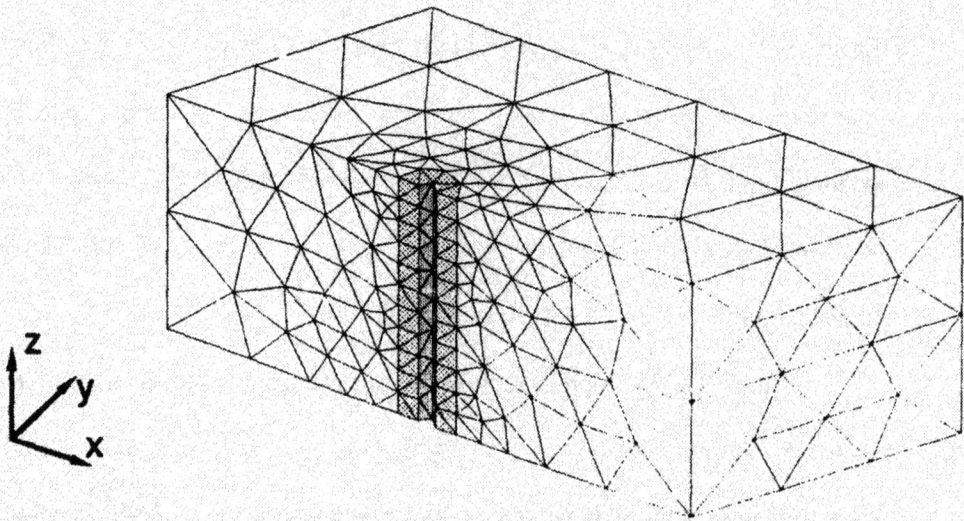

Figure 49. Finite Element Mesh for a Clad Monopole in a Triangular Lattice

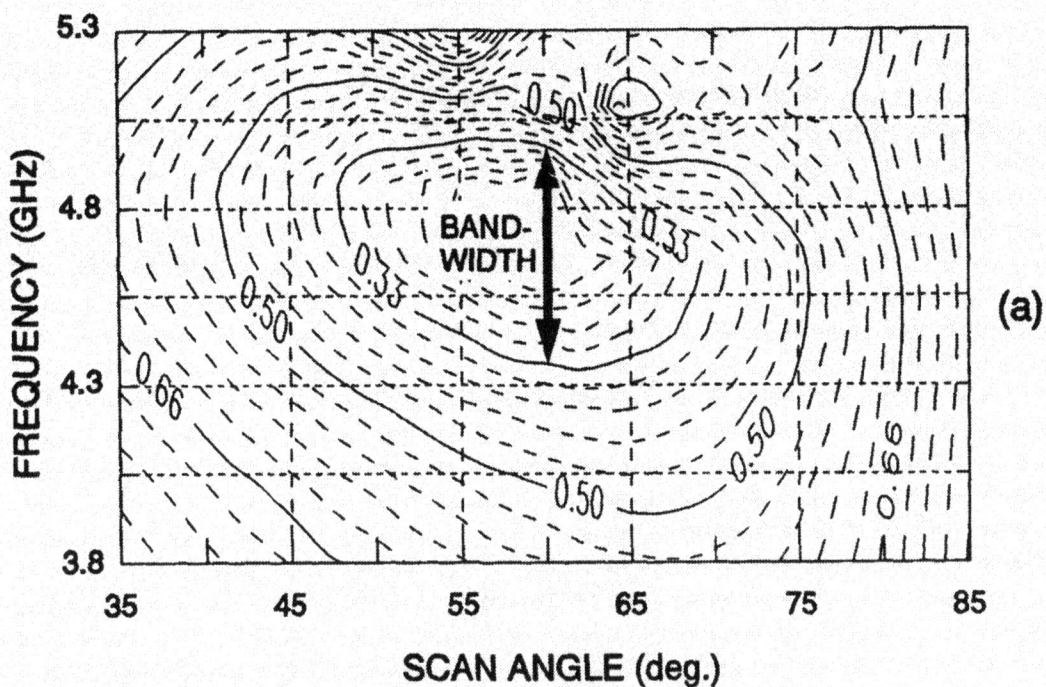

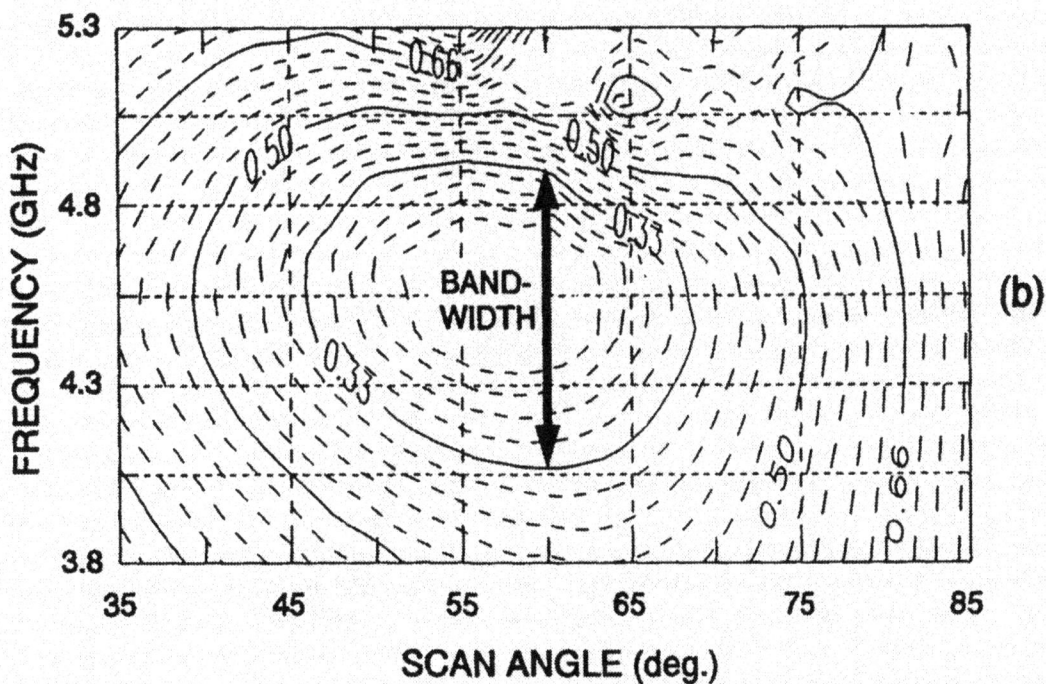

Figure 50. Active Reflection Coefficient Magnitude vs. Scan Angle (horizontal axis, deg.) and Frequency (vertical axis, GHz) for Bare (top) and Clad (bottom) Monopoles

$\theta=60°$. The clad monopole array appears to have nearly twice the bandwidth of the unclad array at $\theta=60°$. This simple example case demonstrates the usefulness of this code for investigating radiator designs - the improved (clad) radiator could not be simulated with existing method of moments codes.

9.4.3. Experiment. A 121-element clad monopole array was fabricated using the same lattice parameters ($d_x=36$mm, $d_y=31$mm) in an isosceles lattice. The monopole element simply an unmodified SMA connector. When installed in the 1/8"-thick Aluminum backplane, the conducting monopole and dielectric cladding extend 14.7mm and 11.8mm, respectively, above the ground plane.

The array layout is shown in Figure 51. The measurement procedure consists of connect-

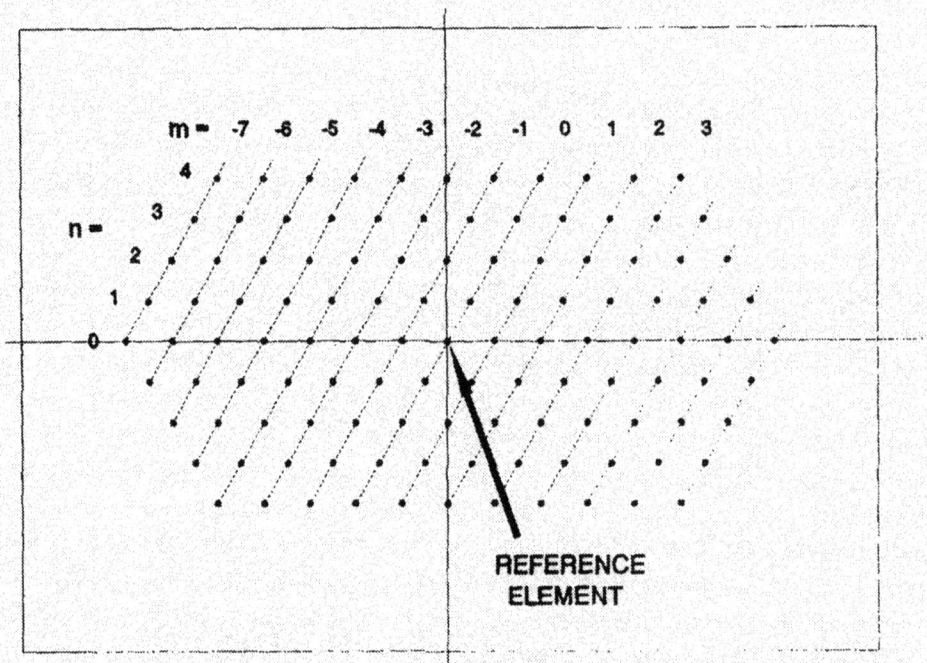

Figure 51. Experimental Array Geometry used for Coupling Measurements

ing the center element to port #1 of a network analyzer, then connecting each other element in turn to port #2 and measuring the coupling coefficient C_{mn}.

The active reflection coefficient as a function of angle is given by

$$R_a(\theta, \phi) \approx \sum_{m=-\infty}^{\infty} \sum_{n=-\infty}^{\infty} C_{mn} e^{-j\psi_x (md_x + nd_y \cot\gamma)} e^{-j\psi_y nd_y} \qquad (66)$$

Once a finite number of C_{mn}'s are found by measurement, an approximation to R_a may be computed for all angles. Figure 52 compares measured results and PARANA calculations. The oscillation (with angle) of the measured data about the actual R_a was expected due to the finite array size.

These results, combined with the earlier waveguide and microstrip patch results provided the validation for the PARANA code. The final two radiator designs discussed next are attempts to exploit the code to assess the properties of radiators for which results are not available by other computational methods.

9.5. Printed Dipole Radiator

9.5.1. Element Design. The design for the dipole element shown earlier in Figure 2 follows general guidelines given by Edward & Rees [6]. Its height is $.25\lambda_0$; its overall length is $.5\lambda_0$ and its arms are $.05\lambda_0$ wide. The design center frequency is 4.8 GHz, giving $\lambda_0 = 62.5$mm. The substrate material chosen for this case is 50 mils thick with relative permittivity of 10 because such material is readily available and it provides a good scaled representation of 100μm GaAs (Gallium Arsenide) at millimeter wave frequencies.

The actual dimensions used for this test case are shown in Figure 53. Figure 54 is an exploded view of the tetrahedron mesh used as the input to PARANA, showing the three mesh regions corresponding to the substrate and the two air regions on each side filling the unit cell.

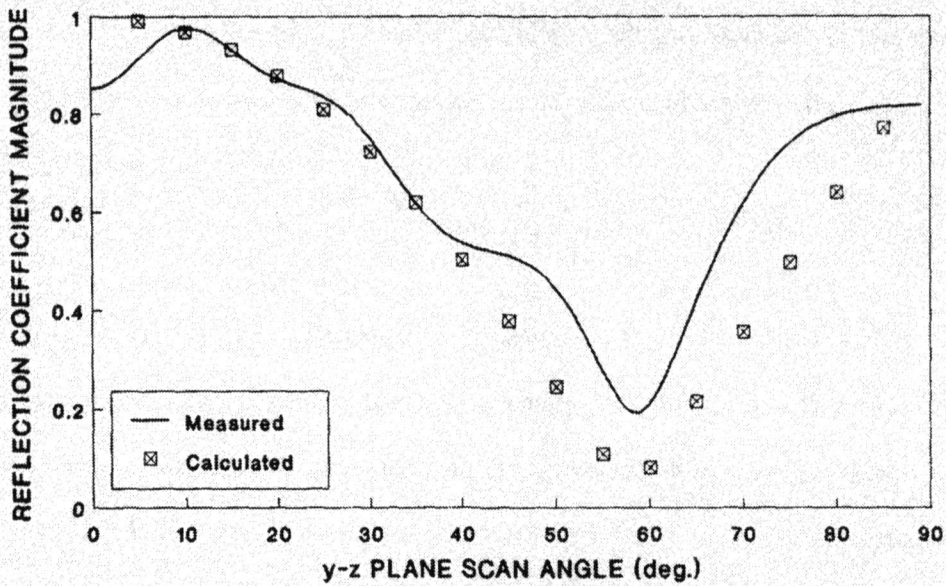

Figure 52. Measured and Computed Active Reflection Coefficient vs. Angle
for Clad Monopole Array Experiment

The array lattice is square, with $.5\lambda_0$ inter-element spacing. Note that the mesh is denser in the dielectric slab (by a factor of $\epsilon_r^{1/2}$), and gradually relaxes going out towards the sides of the unit cell. The dipole and balun are metallized (or photoetched) on opposite sides of the substrate. The slot in the dipole center divides that part of the structure into two coupled microstrip lines. They are to be driven 180° out of phase by the balun. The dipole width is chosen as three times the balun width so that it provides a ground plane for the microstrip lines comprising the balun. In the design shown in Figure 53, the first arm of the balun is a linear taper from microstrip line widths corresponding to 50Ω (the same as the coaxial input) to 80Ω (the narrow end). The second and third balun arms must have the same characteristic impedance as the coupled microstrips, and values below 80Ω require extremely narrow slots that are difficult to construct given the tolerances of photolithographic processes. The slot length is approximately 1/4 guide wavelength from its closed end to where it crosses underneath the microstrip line. Similarly, the microstrip line

84

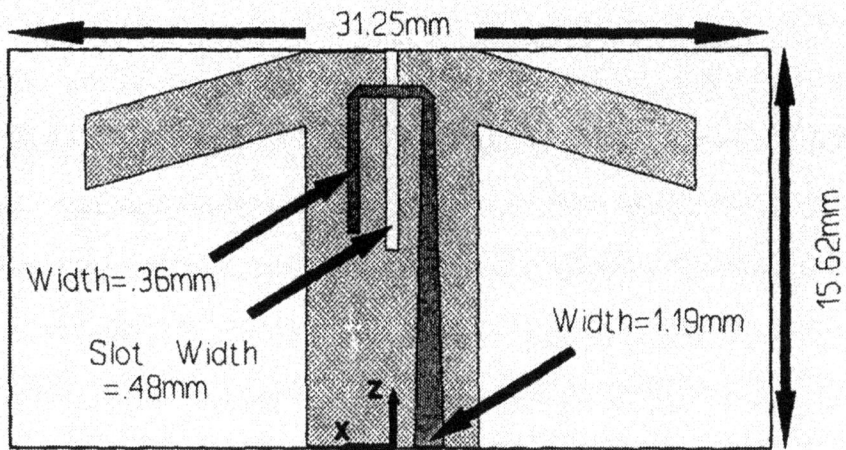

Figure 53. Printed Dipole Radiator Design

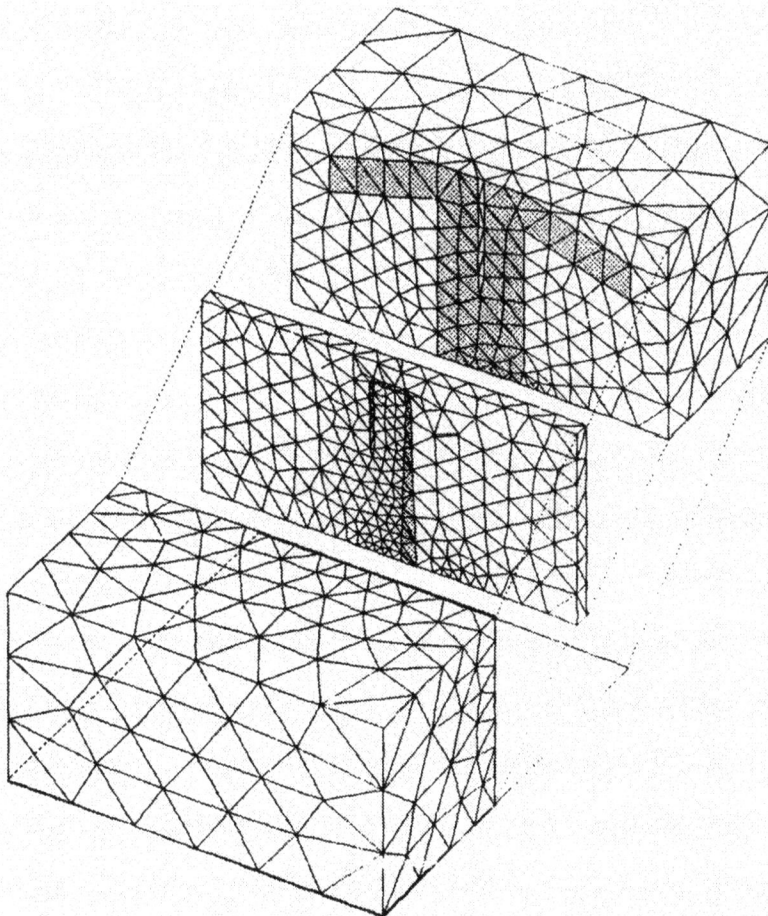

Figure 54. Exploded Finite Element Mesh for Printed Dipole Radiator

length is 1/4 guide wavelength from the slot to its open end, but it is reduced by an effective length using an approximate formula due to Hammerstad [48] (see also [49:190]).

9.5.2. *Calculations.* Initial calculations of R_a vs. frequency revealed that the design in Figure 53 was a poor radiator, with $|R_a|$ greater than .8 at the design center frequency (4.8 GHz). Slightly better results were obtained by scaling the height by .8 so that the overall height was 12.5mm ($.20\lambda_0$) and then reducing the dipole width from $.4\lambda_0$ to $.33\lambda_0$. Figure 55 shows $|R_a|$ vs. frequency for both dipole widths with the 12.5mm height. Unfortunately, this printed dipole has not been tested as an isolated radiator on high-permittivity substrates, so it is not known how much of the mismatch is due to array effects vs. feed line mismatches. Nonetheless, the scanning properties may still be evaluated, normalizing the active reflection coefficient using (65). The results for 4.8 GHz, shown in Figure 56, indicate that there are no scan blindnesses in either the E- or H-plane (the E plane is the plane containing the substrate).

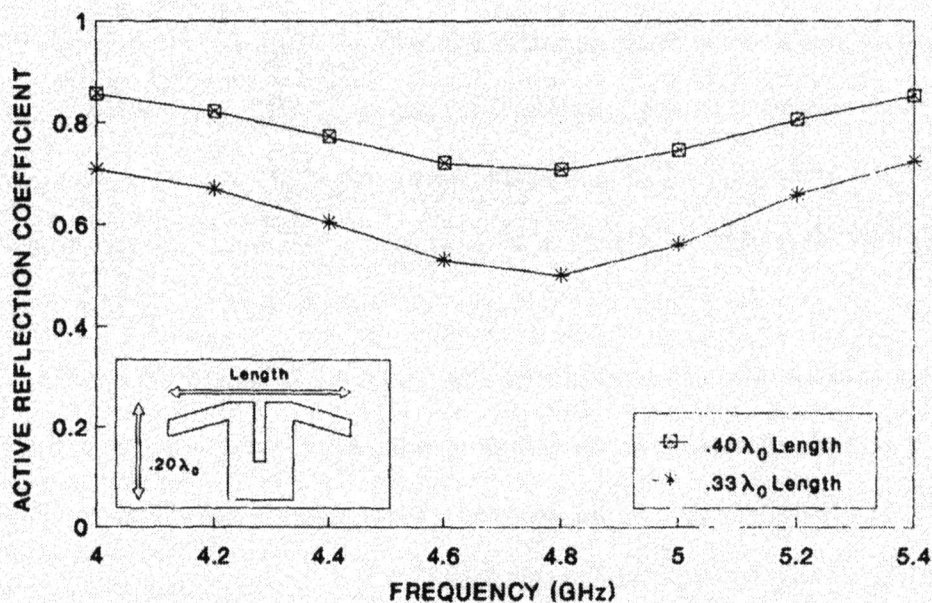

Figure 55. Computed Active Reflection Coefficient at Broadside Scan for Reduced-Height (12.5mm) Printed Dipole for Two Dipole Lengths

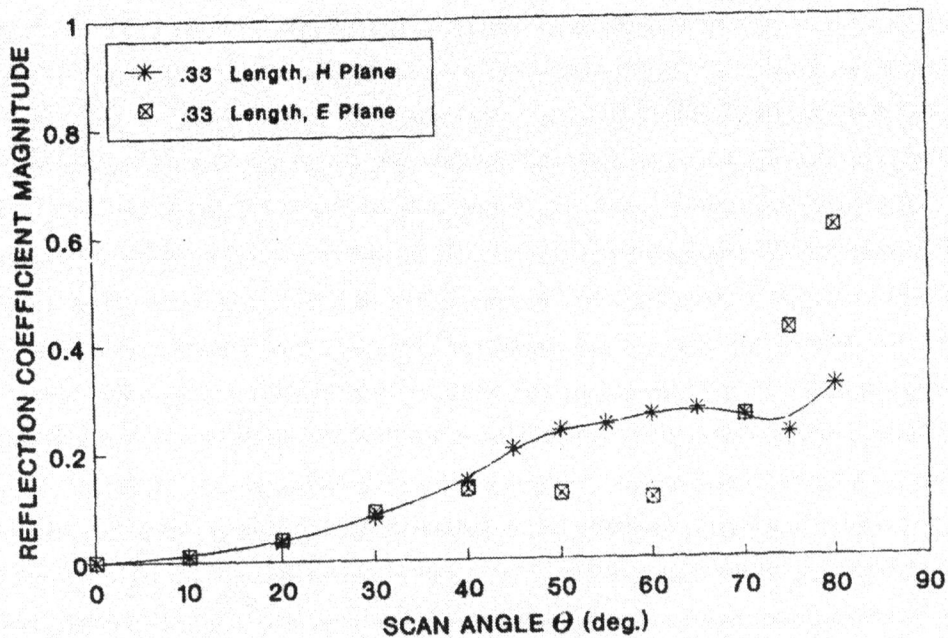

Figure 56. Computed Active Reflection Coefficient (Normalized) vs. Scan Angle
for Reduced Height (12.5mm) Printed Dipole, 4.8 GHz

9.6. Flared Notch Radiator.

9.6.1. Element Design.
The basic idea behind the flared notch radiator shown earlier

in Figure 1 is a slotline, gradually opening out to provide a tapered impedance match to free

space. There do not appear to be any specific design rules [50], but generally, the longer the

flare, the greater the bandwidth. For purposes of this study, the exponential flare given by

Choung & Chen was selected [51].

Figure 1 showed the slotline being fed by a microstrip line, which is in turn fed by a

coaxial cable. This is a form of balun (the same arrangement used for the dipole in the preceding

section) matching the balanced coax to the unbalanced slotline. An alternative balun design is

based on a new coplanar waveguide (CPW)-to-slotline transition that terminates one side of the

CPW in a broadband open circuit [52]. This design has the advantage that only one side of the substrate is metallized, reducing the number of steps in fabrication and eliminating the possibly of registration error. The metallization pattern for a single flared notch radiator is the shaded area in Figure 57. The test case used a 50-mil thick substrate with relative permittivity of 6.0. The flare length and mouth width are 33.3mm and 30.0mm, respectively. The flare shape is given by

$$w(z) = w_0 \exp\left\{\frac{z}{L} \ln \left[\frac{w_m}{w_0}\right]\right\} \tag{67}$$

where w_0 and w_m are the widths at the slotline an the mouth and L is the flare length.

9.6.2. *Array Performance Calculations.* One of the geometry models used as the input to PARANA is shown in Figure 58. As was the case with the dipole element, the mesh is denser in the substrate than in the free space regions. The unit cell size is $d_x = 36$mm and $d_y = 34$mm (rectangular lattice). A second model (not shown) used an equilateral triangular lattice with $d_x = 62$mm, $d_y = 36$mm and $\gamma = 60°$. Both of them used feed waveguide dimensions corresponding

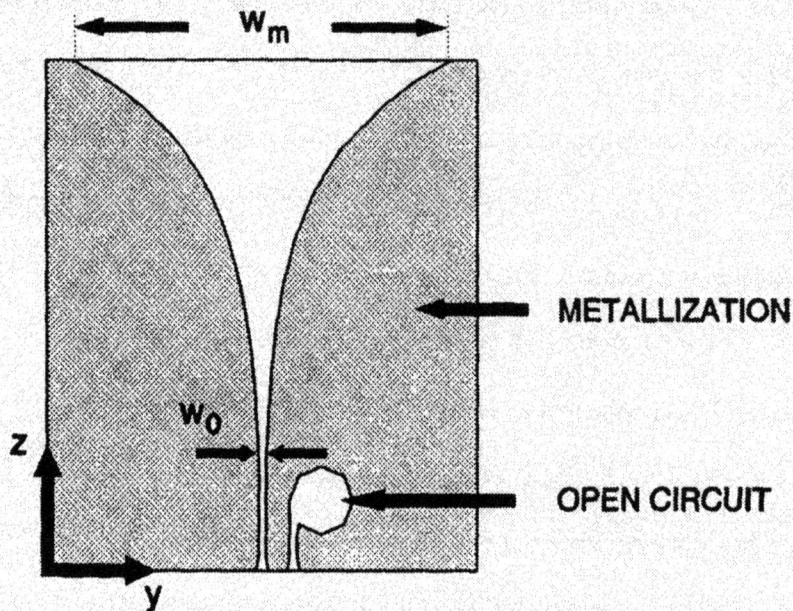

Figure 57. Flared Notch Element Design

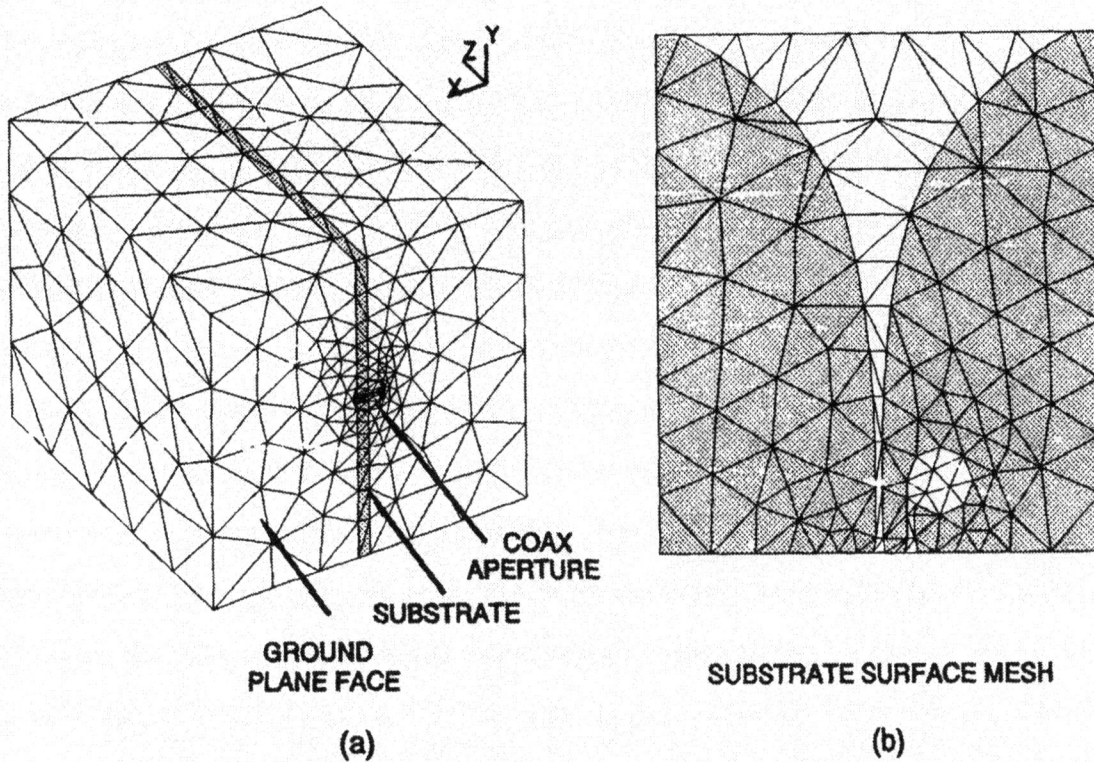

Figure 58. Finite Element Mesh for Flared Notch Radiator (Rectangular Lattice):
(a) Unit Cell Showing Coax Aperture; (b) Substrate Surface Mesh

to APC-3.5 coax (a=.75mm, b=1.75mm).

The broadside ($\theta_0=0$) active reflection coefficient is shown as a function of frequency in Figure 59. These indicate that the radiator is capable of very broad bandwidth, but the skewed-lattice array has a resonance effect that causes a blindness near 4.25 GHz. The rectangular-lattice array is a very promising design, since its predicted bandwidth of greater than 50% is difficult to achieve in an array.

The active reflection coefficient vs. scan angle for the rectangular-lattice array is shown in Figure 60. From the low (3GHz) to the center (4GHz) of the frequency range, it behaves well, but at the high end (5GHz) it displays blindnesses in both scan planes (the E plane is the plane containing the substrate).

89

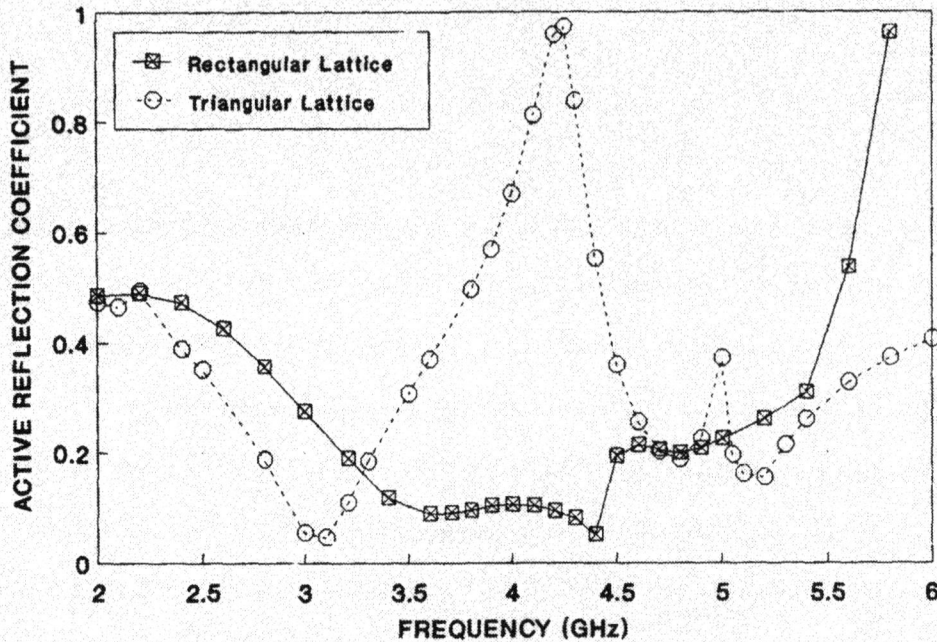

Figure 59. Active Reflection Coefficient vs. Frequency for Printed Flared Notch Arrays

9.7. Summary

The results of this chapter have proven the validity of the essential feature that makes the hybrid finite element method applicable to infinite array analysis: the periodic boundary condition implemented by "wrapping" opposing unit cell edges onto each other with an appropriate phase shift. It was successful for both rectangular and trapezoidal unit cells, the latter applying to skewed array lattices. The microstrip patch array test case showed that it correctly predicts the behavior of surfaces waves in a dielectric layer on a ground plane. The monopole array test case further demonstrated the method's ability to deal with inhomogeneous dielectrics. Most importantly, the same computer program with no changes whatsoever executed the computations for every one of the seven separate array/radiator geometries discussed in this chapter. The only things that changed were the finite element model created in I-DEAS™ and the user instructions

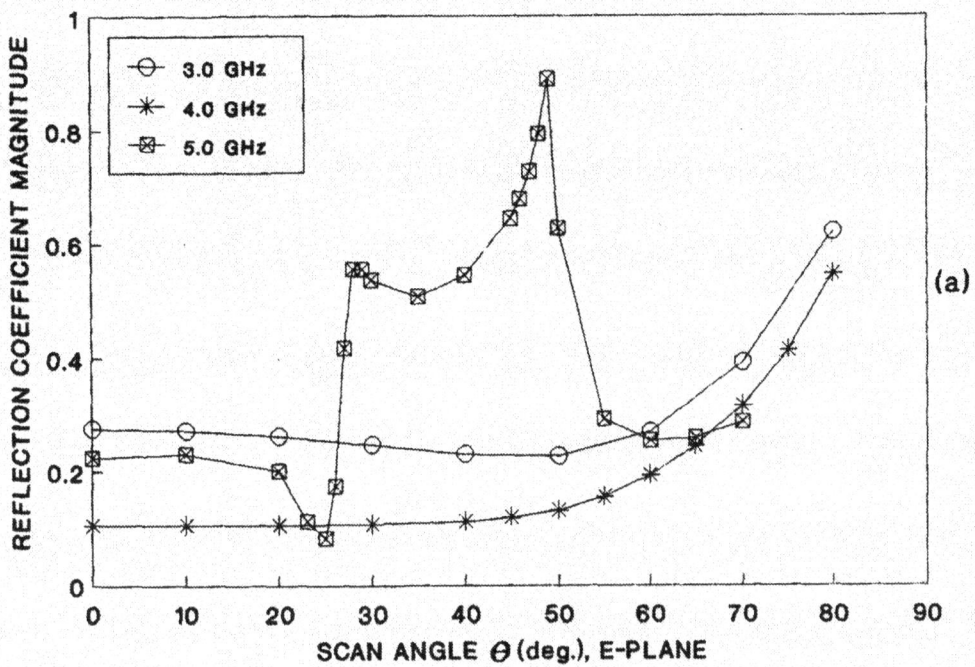

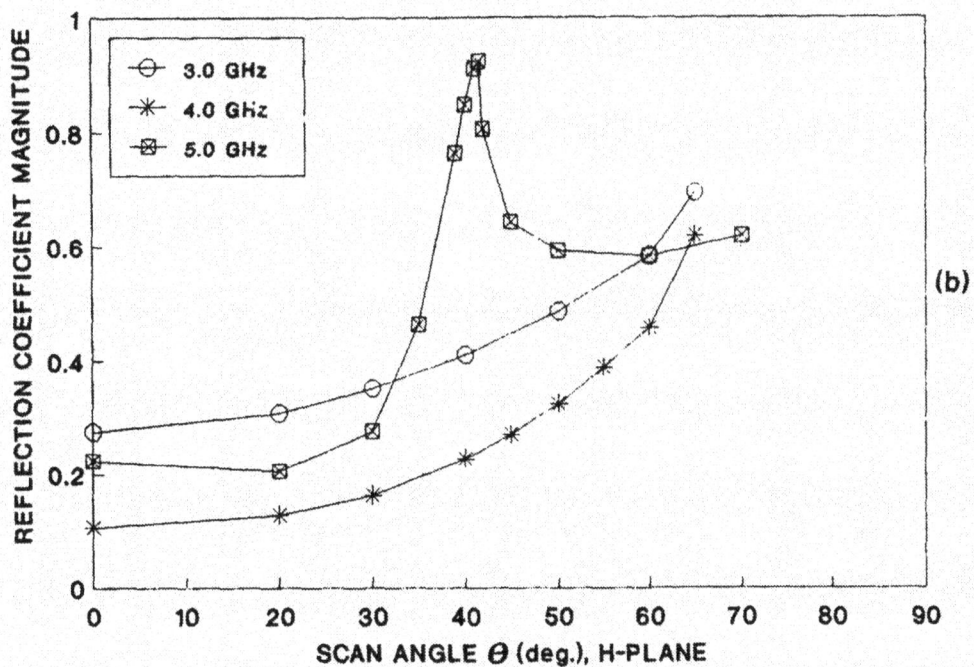

Figure 60. Active Reflection Coefficient vs. Scan Angle, Printed Flared Notch
in Rectangular Lattice: (a) E-Plane Scan; (b) H-Plane Scan

identifying the feed waveguide type and location and the array lattice parameters. Thus it is demonstrated that the finite element method has unparalleled versatility for phased array analysis.

The question of efficiency is addressed below in Table II, which summarizes the time and storage requirements for most of the test cases discussed in this chapter. The total run time (tabulated figures are for a VAX® 4400 mincomputer whose performance is rated at about 17 MIPS) is dominated by the matrix solve time, which was as high as 7 hours per point for the printed dipole (the most time-consuming case). The most time-consuming part of the matrix fill is the calculation of the radiating aperture terms. It is highest for the microstrip patch because that case had relatively more edges in the aperture than did the other cases. The fact that these computations could be performed on a typical minicomputer are encouraging, although whether or not one regards them as "efficient" depends on the difficulty of solving the design problem by other means.

Table II. Mesh Size, Execution Time and Matrix Storage for PARANA Test Cases

Test Case	Unit Cell Vol. (λ^3)	Mesh Cells	Mesh Edges	Iter-ations	Fill Time (min.)	Solve Time (min.)	Matrix Size (Mbytes)
Microstrip Patch	.015	4291	4862	2500	44.5	198	6.2
Clad Monopole	.063	3092	3487	19000	7.1	201	0.7
Flared Notch with Triangular Lattice	.17	4443	4778	18000	8.7	124	0.8
Flared Notch with Rectangular Lattice	.19	4845	5271	23500	9.5	196	0.9
Printed Dipole	.05	6659	7447	39000	20.6	428	1.6

X. Conclusions and Recommendations

This research project began with a novel concept for modeling infinite phased arrays and concluded with a demonstration of its capability. The work in between involved the entire process of electromagnetic predictive code development: casting the physical problem as a mathematical boundary value problem; mapping the latter to a linear system using finite element and moment methods; designing, writing and troubleshooting the general-purpose computer program; adapting the commercial software for geometry generation; and finally, validating the code. Many difficulties needed to be overcome, both expected and otherwise; yet, other expected problems proved inconsequential. This chapter attempts to summarize those findings and to assess the implications of the results to future work.

10.1. Conclusions

10.1.1. Theory and Formulation.
The successes of other researchers in applying the finite element method to time-harmonic electromagnetic problems was reason to believe that it would also succeed for the phased array problem. Nonetheless there are always doubts surrounding any such implementation given that the problem does not have the properties of self-adjointness or positive definiteness. The attendant risk that the iterative matrix solver might converge to a false solution, or not converge at all, did not materialize. The conclusion is that the weak form and Galerkin's method are appropriate to this class of problems.

There have not been any cases that would indicate instability or non-uniqueness, which are always concerns when the properties of fields in two or more regions must be met. A unique solution generally requires that continuity of tangential electric and magnetic fields must be independently enforced. The fact that this problem involves boundaries that are planar and exterior solutions in terms of discrete modes whose tangential field components are dependent

93

evidently circumvents that requirement.

   10.1.2. Implementation. Three-dimensional finite element problems involve very large systems of equatio. s. As seen repeatedly in the previous four chapters, even relatively small devices result in grids with thousands of edges, usually because of the need to capture fine d tails in the geometry. The uncertainty of whether the available computers could solve the resulting matrices in a reasonable time has been resolved and the validation tests indicate that most practical array radiators may be analyzed using typical minicomputers.

   The three codes TWOPORT, CAVIARK and PARANA are geometry-independent, within the constraints of the generic classes of problems they are designed to solve. They exploit the commercial CAD software that benefits from decades of research oriented towards mechanical engineering applications. Two observations regarding that software are: (a) it is capable of generating grids suitable for phased array analysis; and (b), of much greater significance, it removes the geometry generation from the electromagnetic problem so that the codes may encompass far broader problem classes than previously attei. ,ted.

   10.1.3. Validation. The extensive set of validation cases that were used to test the three computer codes demonstrate the effectiveness of the key elements of the solution approach. The interior finite element solution accurately incorporates electromagnetic boundary conditions at perfect conductors (including sharp edges) and dielectric interfaces. It is evidently free of spurious, non-physical solutions, indicating that the divergence condition is also satisfied.

   The waveguide mode integral equation was shown to be an effective mechanism for enforcing field continuity at waveguide apertures. Its payoff is in providing an accurate means for modeling antenna feed structures.

   The hybridization of the periodic integral equation with the finite element solution was shown to correctly enforce the radiation condition above the infinite array. Its implementation ion in the CAVIARR code allowed the correct prediction of active reflection coefficient for a variety of radiator types, including cases involving scan blindnesses.

The implementation of a periodic boundary condition on a three-dimensional finite element problem is evidently a first. The success of the final stage of this work hinged entirely on that single unproven algorithm and the question of whether or not it would conflict with the periodic integral equation. Thus, the most important finding is that the algorithm involving boundary "wrapping" works as proposed, and complements the periodic radiation condition.

10.2. Recommendations

The recommendations fall into three broad categories. The first deals with hardware experiments needed to demonstrate the potential for *design*, in contrast to the *analysis* that was the main subject of this project. The initial designs for the printed dipole and flared notch radiators were a first cut, and their performance leaves much to be desired. Further experiments will be necessary to: first, perfect the design for single, isolated radiators; second, identify the geometry parameters that influence their impedance match in the array environment; and finally, use the codes to optimize those parameters for best performance over some specified range of scan angles and frequency.

The second category deals with possible improvements and enhancements to the computer codes. For example, the TWOPORT code could easily be extended to deal with multi-port RF devices. The inclusion of that feature in PARANA would allow the simulation of radiators that have more than one feed port, such as dual-polarized and multiple-frequency antennas. All three codes would benefit from faster matrix solution, which could result from improved iterative methods, perhaps using preconditioning.

The third and final category of recommendations deals with related or similar electromag-

95

netic problems that may benefit from application of hybrid finite element methods. One that is a particularly straightforward extension of the present work is gratings and frequency selective surfaces. That problem may be addressed simply by replacing the waveguide feed with a second periodic radiation condition and an incident field in the form of a plane wave. A more difficult extension, but one with many practical applications, is a finite-by-infinite array. Such a model could be used to predict the radiating properties of line-source arrays or to assess edge effects in planar arrays. Last, the calculation of scattering from objects that include cavities, or of coupling into circuitry inside those cavities are problems that may benefit from the use of the finite element method to model the cavity interior, with an integral equation boundary condition at the aperture. These problems are more easily addressed now because the present work has provided, among other things, a body of well-tested computer routines for three-dimensional finite element and waveguide mode computations.

Appendix A

The Electric Field Functional

A.1. Variational Principle vs. Weak Form

Publications dealing with electromagnetic finite element applications usually begin with one of two forms of *functionals*[2] with little or no discussion as to why one is used and not the other. This appendix discusses their origins and the circumstances in which each is appropriate. The conclusion is that the results of the two methods are indistinguishable for typical electromagnetic radiation and scattering problems.

The first alternative is

$$F_v(\bar{E}) = \frac{1}{2}\iiint\limits_{\Omega}\left[\frac{1}{\mu_r}\nabla\times\bar{E}\cdot\nabla\times\bar{E} - k_0^2\,\epsilon_r\,\bar{E}\cdot\bar{E}\right]dv + jk_0\eta_0\iint\limits_{\Gamma}(\bar{E}\times\bar{H})\cdot\hat{n}\,ds \qquad (A.1)$$

(see, for example, Jin & Volakis [53]) where Ω is the volume region over which the unknown field is to be found and Γ is its enclosing boundary. The subscript v denotes that this functional is the *variational principle*.[3] The second alternative is

$$F_w(\bar{E}) = \iiint\limits_{\Omega}\left[\frac{1}{\mu_r}\nabla\times\bar{W}\cdot\nabla\times\bar{E} - k_0^2\,\epsilon_r\,\bar{W}\cdot\bar{E}\right]dv + jk_0\eta_0\iint\limits_{\Gamma}(\bar{W}\times\bar{H})\cdot\hat{n}\,ds \qquad (A.2)$$

(see, for example, D'Angelo & Mayergoyz [24]). Here, $\bar{W}$ is a trial function whose form is yet to be determined. The subscript w on F denotes the fact that this is the *weak form*.

---

[2] A functional is a mapping from a space of functions to the complex numbers. It is usually an integral containing an unknown function. The result of the integration is a single number, in contrast to an integral equation, which maps the function into another function.

[3] A *variational statement* may be either a *variational principle* or a *weak form*.

The variational principle F_v has been deduced from a general from of energy functional [54] and is used in conjunction with the Rayleigh-Ritz principle in order to form a system of equations. The weak form is derived more directly by simply taking the inner product of the operator equation with the trial function. It will be used in conjunction with the method of weighted residuals to form the system of equations. In the case of Galerkin's method (a specialization of weighted residuals) the two forms may give exactly the same system of equations. But in order to use Galerkin's method, the expansion function that is admissible in the original problem must also be admissible in the *adjoint problem*.

The following section will discuss the meaning of the adjoint problem and what the conditions are for self-adjointness as applied to the vector wave equation. The third will then show that typical waveguide continuity conditions represent non-self-adjoint boundary conditions. The last section will show that under the assumption that Galerkin's method is applicable, the two formulations generate identical systems of equations.

A.2. The Adjoint Problem

Before deciding whether to use (A.1) or (A.2) one must know whether or not the operator equation is self-adjoint. If not, then the properties of the adjoint problem must be determined in order to ensure that the trial functions are capable of representing its solutions.

The operator equation is the time-harmonic, source-free vector wave equation for the electric field in a linear, isotropic, inhomogeneous region:

$$L(\bar{E}) = \nabla \times \frac{1}{\mu_r} \nabla \times \bar{E} - k_0^2 \epsilon_r \bar{E} = 0 \qquad (A.3)$$

Its inner product with an arbitrary complex function $\bar{W}$ is

$$\langle L(\bar{E}), \bar{W} \rangle = \iiint_{\Omega} \left[\nabla \times \frac{1}{\mu_r} \nabla \times \bar{E} - k_0^2 \epsilon_r \bar{E} \right] \cdot \bar{W}^* \, dv = 0 \qquad (A.4)$$

Using a Green's identity (integration by parts) twice shifts both derivatives from $\bar{E}$ to $\bar{W}$:

$$\langle L(\bar{E}), \bar{W} \rangle = \iiint_{\Omega} \left[\frac{1}{\mu_r} \nabla \times \bar{E} \cdot \nabla \times \bar{W}^* - k_0^2 \epsilon_r \bar{E} \cdot \bar{W}^* \right] dv$$
$$- \iint_{\Gamma} \frac{1}{\mu_r} \bar{W}^* \times \nabla \times \bar{E} \cdot \hat{n} ds \qquad (A.5)$$

$$= \iiint_{\Omega} \bar{E} \cdot \left[\nabla \times \frac{1}{\mu_r^*} \nabla \times \bar{W} - k_0^2 \epsilon_r^* \bar{W} \right]^* dv$$
$$- \iint_{\Gamma} \frac{1}{\mu_r} \left[\bar{W}^* \times \nabla \times \bar{E} - \bar{E} \times \nabla \times \bar{W}^* \right] \cdot \hat{n} ds \qquad (A.6)$$

From the definition

$$\langle L(\bar{E}), \bar{W} \rangle = \langle \bar{E}, L^a(\bar{W}) \rangle \qquad (A.7)$$

it is evident that the term in brackets in the first integral of (A.6) is L^a, the *adjoint operator*. It is simply the wave equation with the constitutive parameters replaced by their complex conjugates. It is now clear that $\bar{W}$ must be in the domain of the adjoint operator.

The definition of self-adjointness is $L=L^a$, which obviously cannot be true if the problem includes lossy materials. But it also depends on whether the boundary conditions are such as to make the surface integral in (A.6) vanish, i.e.

$$\iint_{\Gamma} \frac{1}{\mu_r} \bar{E}^{a*} \times \nabla \times \bar{E} \cdot \hat{n} ds = \iint_{\Gamma} \frac{1}{\mu_r} \bar{E} \times \nabla \times \bar{E}^{a*} \cdot \hat{n} ds \qquad (A.8)$$

$$\iint_{\Gamma} \frac{\mu_r}{\mu_r^*} \bar{E}^{a*} \times \bar{H} \cdot \hat{n} ds = \iint_{\Gamma} \frac{\mu_r^*}{\mu_r} \bar{E} \times \bar{H}^{a*} \cdot \hat{n} ds \qquad (A.9)$$

99

Suppose that all lossy magnetic materials are confined inside Ω so that along the boundary Γ, the permeability is entirely real. Then (A.9) is satisfied when $\bar{E}^a = \bar{E}^*$ and $\bar{H}^a = \bar{H}^*$. Recall that the time-harmonic and time-dependent fields are related by [55:15]

$$\bar{E}(x,y,z,t) = \sqrt{2}\, \text{Re}(\bar{E}e^{j\omega t}) \tag{A.10}$$

(boldface represents the time-dependent quantity; the expression for magnetic field is identical). Evaluating (A.10) at any point $\bar{r}$ along Γ, with $\bar{E}_0$ denoting $\bar{E}(\bar{r})$:

$$\bar{E}(\bar{r},t) = \sqrt{2}\left[\text{Re}(\bar{E}_0)\cos\omega t - \text{Im}(\bar{E}_0)\sin\omega t\right] \tag{A.11}$$

and assuming that $\bar{E}^a = \bar{E}^*$

$$\begin{aligned}
\bar{E}^a(\bar{r},t) &= \sqrt{2}\,\text{Re}\left[\bar{E}_0^* \, e^{j\omega t}\right] = \sqrt{2}\left[\text{Re}(\bar{E}_0)\cos\omega t + \text{Im}(\bar{E}_0)\sin\omega t\right] \\
&= \sqrt{2}\,\text{Re}\left[\bar{E}_0 e^{-j\omega t}\right] \\
&= \bar{E}(\bar{r},-t)
\end{aligned} \tag{A.12}$$

In other words, the adjoint fields are time-reversed versions of the original fields. They carry power across the boundary in the opposite direction and they encounter materials that have gain instead of loss. This is the physical interpretation of the adjoint problem, and is consistent with the property that if the operator L is *causal*, then the operator L^a is *anti-causal* [56:356].

Notice that if the boundary Γ is comprised entirely of perfect electric and perfect magnetic conductors, then (A.9) is always satisfied because there cannot be any transfer of power across such boundaries. This leads to the suspicion that the boundary conditions that will cause non-self-adjointness are those open boundaries where conditions of field continuity are to be enforced. Section A.4 will demonstrate that this is indeed the case for the waveguide/cavity apertures that are considered in the main body of this report.

100

A.3. Continuity Conditions for Waveguide Apertures

For the waveguide continuity problem, the boundary Γ reduces to the waveguide aperture Γ_A. Taking $\bar{E}^a = \bar{E}^*$ and $\bar{H}^a = \bar{H}^*$, the following form is equivalent to (A.9):

$$\iint_{\Gamma_A} \bar{E}^* \cdot (\hat{n} \times \bar{H}) ds = \iint_{\Gamma_A} \bar{E} \cdot (\hat{n} \times \bar{H}^*) ds \tag{A.13}$$

(again assuming that μ_r is real along Γ). Consider a waveguide whose axis is the z axis and which joins the volume Ω through an aperture in its end wall at $z=0$. It is assumed to be match-terminated at $z<<0$. The aperture field due to the dominant mode and its conjugate are

$$\begin{aligned} \bar{E} &= \bar{g}_0 (1 + C_0) \\ \bar{E}^* &= \bar{g}_0 (1 + C_0^*) \end{aligned} \tag{A.14}$$

$$\begin{aligned} \bar{H} &= \hat{z} \times \bar{g}_0 Y_0 (1 - C_0) \\ \bar{H}^* &= \hat{z} \times \bar{g}_0 Y_0 (1 - C_0^*) \end{aligned} \tag{A.15}$$

where $\bar{g}_0$ is a transverse mode function, Y_0 is the modal admittance and C_0 is an unknown coefficient. The outward normal to Ω is $\hat{n} = -\hat{z}$, so the left and right sides of (A.13) give

$$\iint_{\Gamma_A} \bar{E}^* \cdot (\hat{n} \times \bar{H}) ds = Y_0 (1 + C_0^*)(1 - C_0) \iint_{\Gamma_A} |\bar{g}_0|^2 ds \tag{A.16}$$

$$\iint_{\Gamma_A} \bar{E} \cdot (\hat{n} \times \bar{H}^*) ds = Y_0 (1 + C_0)(1 - C_0^*) \iint_{\Gamma_A} |\bar{g}_0|^2 ds \tag{A.17}$$

These two are only equal if C_0 is real, which is not generally the case. Therefore, the continuity condition across a waveguide aperture will render the boundary value problem non self-adjoint; the functional (A.1) is not appropriate even when there are no lossy materials; and the weak form

101

(A.2) should be chosen.

A.4. Galerkin's Method vs. Rayleigh-Ritz

In the main body of this report, the finite element method is used to produce a system of equations from the weak form functional. The field is represented by a summation of unknown complex coefficients with known, linear vector functions:

$$\bar{E} \approx \tilde{E} = \sum_{s=1}^{N} e_s \bar{\psi}_s (x,y,z) \tag{A.18}$$

For the expansion functions to be *admissible*, they must be in the domain of the functional. Linear functions meet that requirement since their first derivatives are continuous (integrable).

The residual error is defined as $\bar{R} = L(\bar{E}) - L(\tilde{E})$. N weighting functions will be chosen, each of which is orthogonal to the residual so that $\langle \bar{R}, \bar{w}_r \rangle = 0$, giving the system of N equations

$$0 = \sum_{s=1}^{N} e_s \iiint_{\Omega} \left[\frac{1}{\mu_r} \nabla \times \bar{w}_r \cdot \nabla \times \bar{\psi}_s - k_0^2 \epsilon_r \bar{w}_r \cdot \bar{\psi}_s \right] dv$$
$$+ jk_0 \eta_0 \iint_{\Gamma} (\bar{w}_r \times \bar{H}) \cdot \hat{n} \, ds, \quad r = 1,2,...,N \tag{A.19}$$

The choice $\bar{w}_r = \bar{\psi}_r$ satisfies the orthogonality requirement, but it is also required that $\bar{\psi}_r$ be an admissible expansion function for the adjoint electric field. Starting with the adjoint operator equation, forming $\langle \bar{E}, L^a(\bar{W}) \rangle$ and using the Green's identity once gives

$$\langle \bar{E}, L^a(\bar{W}) \rangle = \iiint_{\Omega} \left[\frac{1}{\mu_r} \nabla \times \bar{W}^* \cdot \nabla \times \bar{E} - k_0^2 \epsilon_r \bar{W}^* \cdot \bar{E} \right] dv$$
$$- \iint_{\Gamma} \frac{1}{\mu_r} \bar{E} \times \nabla \times \bar{W}^* \cdot \hat{n} \, ds \tag{A.20}$$

which makes it evident that expansion functions for $\bar{W}$ must have continuous first derivatives. Therefore, the same expansion functions are admissible in both the original and the adjoint problems. It is this fact that allows Galerkin's method to be employed.

Consider the variational functional (A.1) with the expansion (A.18) for the electric field:

$$
F_v(\bar{E}) = \frac{1}{2} \sum_{r=1}^{N} e_r \sum_{s=1}^{N} e_s \iiint_{\Omega} \left[\frac{1}{\mu_r} \nabla \times \bar{\psi}_r \cdot \nabla \times \bar{\psi}_s - k_0^2 \epsilon_r \bar{\psi}_r \cdot \bar{\psi}_s \right] dv
$$
$$
+ jk_0 \eta_0 \sum_{r=1}^{N} e_r \iint_{\Gamma} (\bar{\psi}_r \times \bar{H}) \cdot \hat{n} \, ds, \quad r = 1, 2, \dots, N
$$

(A.21)

The Rayleigh-Ritz method equates the stationary point of F_v to the minimization of the above with respect to each of the coefficients e_r, i.e.

$$
\delta F_v(\bar{E}) = 0 \iff \frac{\partial F_v(\bar{E})}{\partial e_r} = 0, \ \forall r
$$

(A.22)

Carrying out the partial derivatives in (A.21) gives

$$
0 = \sum_{s=1}^{N} e_s \iiint_{\Omega} \left[\frac{1}{\mu_r} \nabla \times \bar{\psi}_r \cdot \nabla \times \bar{\psi}_s - k_0^2 \epsilon_r \bar{\psi}_r \cdot \bar{\psi}_s \right] dv
$$
$$
+ jk_0 \eta_0 \iint_{\Gamma} (\bar{\psi}_r \times \bar{H}) \cdot \hat{n} \, ds, \quad r = 1, 2, \dots, N
$$

(A.23)

which is identical to (A.19) with $\bar{w}_r$ replaced by $\bar{\psi}_r$. This shows that the systems of equations resulting from the variational principle and from the weak form are identical. Thus, under at least some circumstances the distinction between the two forms is inconsequential, and the variational principle may also be used.

103

Appendix B

Waveguide Mode Function Inner Products

B.1. Approach

The waveguide interaction terms require the computation of two surface integrals Φ_{si} and Ψ_{si} from (36) and (37) where s and i are the edge and mode indices. Each may be found by summing the contributions from the individual triangles that share edge s, e.g. for triangle k:

$$\Phi_{si}^{(k)} = \frac{L_s}{2A_k}(\hat{n} \cdot \hat{z}) \int_{\Delta_k} \left[g_{ix}(f_2 T_{13} - f_1 T_{23}) + g_{iy}(f_1 T_{22} - f_2 T_{12}) \, dx \, dy \right] \quad \text{(B.1)}$$

L_s is the edge length, $\hat{n}$ is the outward surface normal (into the waveguide), A_k is the triangle area, f_1 and f_2 are the linear scalar finite elements associated with the nodes bounding the edge and T is the 3x3 simplex transformation matrix for the triangle. When the integral is transformed, $dx \, dy \rightarrow 2A_k dt_1 dt_2$ and

$$\Phi_{si}^{(k)} = L_s (\hat{n} \cdot \hat{z}) \left[-T_{23} G_{x1} + T_{13} G_{x2} + T_{22} G_{y1} - T_{12} G_{y2} \right] \quad \text{(B.2)}$$

$$G_{vj} = \int_0^1 dt_1 \int_0^{1-t_1} g_{iv} t_j \, dt_2 \,, \quad v = x,y; j = 1,2 \quad \text{(B.3)}$$

Similarly,

$$\Psi_{si}^{(k)} = L_s \left[T_{22} G_{x1} - T_{12} G_{x2} + T_{23} G_{y1} - T_{13} G_{y2} \right] \quad \text{(B.4)}$$

The generic procedure used to compute the terms of the matrix S^{JE} is outlined in Figure B1. The bulk of the computation is in step 2.b.ii, where the integrals G_{vj} are calculated. In the case of rectangular waveguide, they may be evaluated in closed form. For the other two waveguide

```
FOR EACH APERTURE (A,B):
    FOR EACH MODE (i):
        1. COMPUTE PROPAGATION CONSTANT, MODAL ADMITTANCE
        2. FOR EACH TRIANGLE (k) IN APERTURE:
            a. COMPUTE SIMPLEX TRANSFORMATION
            b. FOR EACH EDGE (s) BORDERING THE TRIANGLE:
                i.   DETERMINE EDGE VECTOR ORIENTATION
                ii.  COMPUTE G_{νj} FOR ν=x,y and j=1,2
                iii. ADD CONTRIBUTIONS Φ_{si}^{(k)} and Ψ_{si}^{(k)} to Φ_{si} and Ψ_{si}
        3. FOR EACH EDGE (s) IN APERTURE:
            a. FOR EACH EDGE (t) IN APERTURE:
                i. ADD [ j k_0 η_0 Y_i Φ_{si} Ψ_{si} ] to S_{st}^{JE}
```

Figure B1. Procedure for Matrix Fill Calculations Involving Waveguide Modes

types, circular and circular coaxial, they are computed numerically using Gaussian quadrature.

The remainder of this appendix discusses the details of those integrations. Expressions for the

mode functions, cutoff wavenumbers and modal admittances may be found in [17] and [28].

B.2. Rectangular Waveguide

The mode functions for rectangular waveguide will have indices m,n and p, where p=1

or 2 for TE or TM, respectively. The vector components are

$$g_{mnpx} = C_{mnpx} \cos\left(\frac{m\pi x}{a}\right) \sin\left(\frac{n\pi y}{b}\right) \qquad (B.5)$$

$$g_{mnpy} = C_{mnpy} \sin\left(\frac{m\pi x}{a}\right) \cos\left(\frac{n\pi y}{b}\right) \qquad (B.6)$$

where a and b are the waveguide dimensions along the x and y axes, respectively. The normaliztion coefficients ensure that the modes are orthonormal over the waveguide cross section.

The inverse of the simplex transform matrix, T^{-1}, will give x and y in terms of t_1, t_2 and constant coefficients:

$$x = \alpha_0 + \alpha_1 t_1 + \alpha_2 t_2 \qquad (B.7)$$

$$y = \beta_0 + \beta_1 t_1 + \beta_2 t_2 \qquad (B.8)$$

Let ξ_i and η_i represent the following combinations of α_i and β_i

$$\xi_i = \frac{m\pi\alpha_i}{a} + \frac{n\pi\beta_i}{b} \qquad (B.9)$$

$$\eta_i = \frac{m\pi\alpha_i}{a} - \frac{n\pi\beta_i}{b} \qquad (B.10)$$

In terms of these, the mode functions may be rewritten:

$$\frac{g_{mnpx}}{C_{mnpx}} = \frac{1}{2}\sin(\xi_0 + \xi_1 t_1 + \xi_2 t_2) - \frac{1}{2}\sin(\eta_0 + \eta_1 t_1 + \eta_2 t_2) \qquad (B.11)$$

$$\frac{g_{mnpy}}{C_{mnpy}} = \frac{1}{2}\sin(\xi_0 + \xi_1 t_1 + \xi_2 t_2) + \frac{1}{2}\sin(\eta_0 + \eta_1 t_1 + \eta_2 t_2) \qquad (B.12)$$

Let H_ξ denote the integral

$$H_\xi = \frac{1}{2}\int_0^1 t_1 dt_1 \int_0^{1-t_1} \sin(\xi_0 + \xi_1 t_1 + \xi_2 t_2) dt_2 \qquad (B.13)$$

H_η is identical with η_i replacing ξ_i everywhere. H_ξ' and H_η' are the same, but with ξ_1 and ξ_2 or η_1 and η_2 reversed. In terms of these, the integrals in (B.3) are

$$G_{x1} = C_{mnpx}(H_\xi - H_\eta) \qquad (B.14)$$

106

$$G_{y1} = C_{mnpy}(H_\xi + H_\eta) \qquad \text{(B.15)}$$

$$G_{x2} = C_{mnpx}\left(H'_\xi - H'_\eta\right) \qquad \text{(B.16)}$$

$$G_{y2} = C_{mnpy}\left(H'_\xi + H'_\eta\right) \qquad \text{(B.17)}$$

Finally H_ξ may be evaluated in closed form, but it must be accomplished separetely for several special cases, as given below:

(a) general case:

$$H_\xi = \frac{1}{2\xi_2}\left\{ \frac{1}{\xi_1^2}\left[\cos(\xi_0 + \xi_1) - \cos\xi_0 + \xi_1\sin(\xi_0 + \xi_1)\right] \right.$$
$$\left. - \frac{1}{(\xi_1 - \xi_2)^2}\left[\cos(\xi_0 + \xi_1) - \cos(\xi_0 + \xi_2) + (\xi_1 - \xi_2)\sin(\xi_0 + \xi_1)\right]\right\} \qquad \text{(B.18)}$$

(b) $\xi_2 \neq 0$, $\xi_1 \neq 0$, $|\xi_2 - \xi_1| \ll 1$ (for small values of $\xi_2 - \xi_1$ (B.18) is numerically unstable, susceptible to large roundoff error and overflow):

$$H_\xi \approx \frac{1}{2\xi_2}\left\{ \xi_1^{-2}\left[\cos(\xi_0 + \xi_1) - \cos\xi_0 + \xi_1\sin(\xi_0 + \xi_1)\right] \right.$$
$$\left. - \frac{1}{2}\cos(\xi_0 + \xi_2) + \frac{1}{3}(\xi_1 - \xi_2)\sin(\xi_0 + \xi_1)\right\} \qquad \text{(B.19)}$$

(c) $|\xi_2| \ll 1$:

$$H_\xi = \frac{1}{4\xi_1^4}\left\{ (4\xi_1 + 6\xi_2 - \xi_1^2\xi_2)\cos(\xi_0) - (4\xi_1 + 6\xi_2)\cos(\xi_0 + \xi_1) \right.$$
$$\left. - (2\xi_1^2 + 4\xi_1\xi_2)\sin(\xi_0) - 2(\xi_1^2 + \xi_1\xi_2)\sin(\xi_0 + \xi_1)\right\} \qquad \text{(B.20)}$$

(d) $|\xi_1| \ll 1$:

$$H_\xi \approx \frac{1}{2\,\xi_2^4} \left\{ \frac{1}{2}\cos\xi_0 - \frac{1}{3}\xi_1 \sin\xi_0 \right.$$

$$\left. - \frac{1}{(\xi_1 - \xi_2)^2}\left[\cos(\xi_0 + \xi_1) - \cos(\xi_0 + \xi_2) + (\xi_1 - \xi_2)\sin(\xi_0 + \xi_1) \right] \right\} \qquad (B.21)$$

(e) $|\xi_2| \ll 1$, $|\xi_1| \ll 1$:

$$H_\xi = \frac{1}{12}\left\{ \sin\xi_0 + (\frac{\xi_1}{2} + \frac{\xi_2}{4})\cos(\xi_0) \right\} \qquad (B.22)$$

Note that the last four are correct in the limits as those quantities specified as much less than unity go to zero. Precedence is given to the five formulas in reverse order, checking for condition (e) first, and executing (a) only when none of the others (d), (c) or (b) are true.

B.3. Gaussian Quadrature Integration

For circular and coaxial mode functions the inner products (B.3) may not be accomplished in closed form, so they have been evaluated numerically using Gaussian quadrature. Quadrature formulas only apply strictly to one dimension, but are easily extended to two dimensional integration over rectangular areas. To use these, the triangle's geometry is transformed to a unit square, using an approach suggested by Stroud & Secrest [57]. The transformation to simplex coordinates mapped an arbitrary triangle into one with vertex coordinates are $(0,0)$, $(0,1)$ and $(1,0)$ in (t_1, t_2) coordinates. The second transformation is given by

$$u_1 = t_1 \; ; \quad u_2 = \begin{cases} \frac{t_2}{(1-t_1)}, & t_1 \neq 1 \\ 1, & t_1 = 1 \end{cases} \qquad (B.23)$$

The Jacobian of this transformation is $(1-t_1)^{-1}$, so that a typical integral term transforms into:

$$G_1 = \int_0^1 t_1 \, dt_1 \int_0^{1-t_1} \bar{g}(t_1,t_2) \, dt_2 = \int_0^1 u_1 \, du \int_0^1 (1-u_2)^2 \, \bar{g}(u_1,u_2) \, du_2 \qquad \text{(B.24)}$$

Let u_k and u_m denote the one-dimensional quadrature sample points along u_1 and u_2, respectively, with w_k and w_m the corresponding weights. Then the integral is approximated by the sum

$$G_1 \approx \sum_{k=1}^{Q} w_k \, u_k \sum_{m=1}^{Q} w_m (1-u_m)^2 \, \bar{g}(u_k,u_m) \qquad \text{(B.25)}$$

where Q is the order of the quadrature formula. See, for example, [57] or [58:887] for tables of weights and quadrature points.

Appendix C

The Periodic Integral Equation

Conventional derivations for the scanning properties of phased array antennas are often given in terms of "Floquet modes," which are essentially plane waves propagating in several discrete directions away from or towards the array. That derivation is an analogue to ordinary waveguide modes, and was developed as convenient means for deriving mode-matching solutions to radiation from waveguide arrays. This appendix presents an alternative derivation using Fourier transforms that may, in some cases, lead to greater insight than the mode-matching solution.

This derivation proceeds from the integral equation for an arbitrary sheet current. It is then specialized to the case of an infinite periodic sheet current through the periodicity condition. Its spatial frequency domain representation is then obtained through straightforward application of Fourier transform theorems and a Fourier-Bessel transform for the free space Green's function. The inverse transform then yields the desired result, which is an expression for the integral equation not as a continuous integral, but as a summation over sample points in spatial frequency. An extension of this derivation for skewed array lattices is also provided. The results of Galerkin's method are specialized to the linear vector finite element functions and analytic expressions for the resulting inner products are given. Finally, the derivation in terms of Floquet modes is shown to provide an identical system of equations.

C.1. The MFIE for Planar Current Sources

The objective of this section is to obtain an integral equation for the fields in the half space above the radiation boundary due to the fields below it. The use of the equivalence principle will simplify the derivation. The boundary, which will be taken to be located at $z=0$, supports equivalent currents $\bar{M}$ and $\bar{J}$. The source of these are the tangential fields just below the boundary:

$$\bar{M} = \hat{z} \times \bar{E}(z=0_-) \tag{C.1}$$

$$\bar{J} = -\hat{z} \times \bar{H}(z=0_-) \tag{C.2}$$

The equivalent problem in the $z > 0$ half space also sees a conducting boundary, but it supports $-\bar{M}$ and $-\bar{J}$. This equivalent current is the source of an electric vector potential, $\bar{F}$:

$$\bar{F}(\bar{r}) = \frac{\epsilon}{4\pi} \int\limits_{-\infty}^{\infty}\int -\bar{M}(\bar{r}) \, G(\bar{r}-\bar{r}') \, dx' \, dy' \tag{C.3}$$

$$\bar{A}(\bar{r}) = \frac{\mu}{4\pi} \int\limits_{-\infty}^{\infty}\int -\bar{J}(\bar{r}) \, G(\bar{r}-\bar{r}') \, dx' \, dy' \tag{C.4}$$

$$G(\bar{r}-\bar{r}') = \frac{e^{-jk|\bar{r}-\bar{r}'|}}{|\bar{r}-\bar{r}'|} \tag{C.5}$$

where $\bar{r}'$ and $\bar{r}$ denote, respectively, source and observer coordinates, and G is the time-harmonic free space Green's function. Here, $\bar{r}'$ is confined to $z=0$, but $\bar{r}$ may be anywhere above $z=0$. The magnetic field at the observation point is [59:36]:

$$\bar{H}(\bar{r}) = -j\omega\bar{F} + \frac{\nabla\nabla \cdot \bar{F}}{j\omega\mu\epsilon} \tag{C.6}$$

An integral equation is obtained by applying a boundary condition to the above radiation integral. Specifically, the total tangential magnetic field at $z=0$ is $\bar{H} = \hat{z}\times\bar{J}$:

$$\hat{n} \times \bar{J} = \frac{1}{j\omega\mu\epsilon} \left\{ \hat{x}\left[k^2 F_x + \frac{\partial^2 F_x}{\partial x^2} + \frac{\partial^2 F_y}{\partial x \partial y} \right] + \hat{y}\left[k^2 F_y + \frac{\partial^2 F_y}{\partial y^2} + \frac{\partial^2 F_x}{\partial x \partial y} \right] \right\} \tag{C.7}$$

111

Note that an electric current source in the z=0 plane produces a magnetic field that is entirely normal to that plane. Hence $\bar{A}(z=0) = \hat{z}A_z$, so $\bar{J}$ does not contribute to $\bar{H}_t(z=0)$ and (7) is a complete expression for the integral equation. The next section will specialize this to the case in which $\bar{M}$ is an infinite periodic source.

C.2. *The Periodic Magnetic Field Integral Equation*

According to Floquet's theorem, the fields and currents anywhere on and above an infinite periodic array must obey the relationship

$$\bar{\Phi}(x+md_x,y+nd_y) = \bar{\Phi}(x,y)\,e^{-j\psi_x md_x}e^{-j\psi_y nd_y} \tag{C.8}$$

where Φ may be any of E,H,M, or F, d_x and d_y are the lattic spacings in x and y and ψ_x and ψ_y are the phase shifts necessary to produce a plane wave propagating in the θ_0,ϕ_0 direction:

$$\psi_x = k\,\sin\theta_0\,\cos\phi_0 \tag{C.9}$$

$$\psi_y = k\,\sin\theta_0\,\sin\phi_0 \tag{C.10}$$

Let $\bar{M}_{uc}(x,y)$ denote a unit cell magnetic current that is equal to the source distribution within the rectangular area $-d_x \leq x \leq d_x$ and $-d_y \leq y \leq d_y$ and zero elsewhere.

Consider an infinite two-dimensional sequence of Dirac delta functions located at lattice points $x=md_x$, $y=nd_y$ for $-\infty < m,n < \infty$. Figure C1 illustrates that the effect of a two-dimensional convolution of this Dirac sequence with the unit cell current distribution is to replicate the current distribution around each of the lattice points. If each impulse is also weighted by the complex exponential representing the beam steering phase, then $\bar{M}$ is:

$$\bar{M}(x,y) = \bar{M}_{uc} * \sum_m \sum_n \delta(x-md_x,y-nd_y)\,e^{-j\psi_x x}e^{-j\psi_y y} \tag{C.11}$$

where * denotes the convolution operation. This is an alternative form of Floquet's theorem.

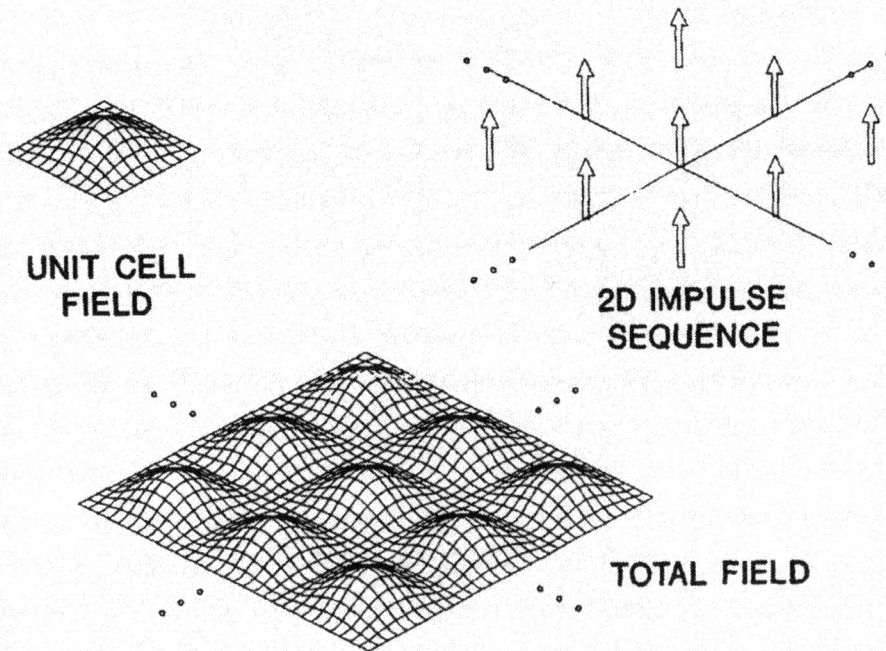

Figure C1. Two Dimensional Convolution of Unit Cell Field with Dirac Impulse Sequence

(All summations may be assumed to have limits of -∞ to +∞ unless otherwise noted.)

Let $\underline{\bar{M}}$ denote the two-dimensional Fourier transform of the magnetic field with respect to the spatial frequency coordinates k_x and k_y (underbar will be used to denote transformed quantities):

$$\underline{\bar{M}}(k_x,k_y) = \int\limits_{-\infty}^{\infty}\!\!\int \bar{M}(x,y)\,e^{jk_x x}\,e^{jk_y y}\,dx\,dy \qquad (C.12)$$

$$\bar{M}(x,y) = \frac{1}{4\pi^2}\int\limits_{-\infty}^{\infty}\!\!\int \underline{\bar{M}}(k_x,k_y)\,e^{-jk_x x}\,e^{-jk_y y}\,dk_x\,dk_y \qquad (C.13)$$

The following four properties of Fourier transforms are required (see, for example [30:199-200]):

$$\mathscr{F}\{F * G\} = \underline{F}\,\underline{G} \qquad (C.14)$$

113

$$\mathcal{F}\{F(x)e^{-j\alpha x}\} = \underline{F}(k_x + \alpha) \tag{C.15}$$

$$\mathcal{F}\left\{\frac{\partial F}{\partial x}\right\} = -jk_x\underline{F} \tag{C.16}$$

$$\mathcal{F}\left\{\sum_{n=-\infty}^{\infty} \delta(x-nd)\right\} = \frac{2\pi}{d}\sum_{n=-\infty}^{\infty} \delta\left(k_x - \frac{2\pi n}{d}\right) \tag{C.17}$$

(The last one is a form of the Poisson sum formula.) Substituting (C.9) into (C.10) and using properties (C.14), (C.15) and (C.17) results in

$$\underline{\bar{M}}(k_x,k_y) = \underline{\bar{M}}_{uc}\frac{4\pi^2}{d_x d_y}\sum_m \sum_n \delta(k_x - k_{xm}, k_y - k_{yn}) \tag{C.18}$$

$$k_{xm} = \frac{2\pi m}{d_x} - \psi_x \tag{C.19}$$

$$k_{yn} = \frac{2\pi n}{d_y} - \psi_y \tag{C.20}$$

The points k_{xm} and k_{yn} are sample points in the spatial frequency, or *spectral* domain. They may also be recognized as the so-called *Floquet harmonics*.

When the source and observer are both in the $z=0$ plane the Green's function is only a function of x and y and the integral (C.3) is a convolution integral written as

$$\bar{F}(x,y) = -\frac{\epsilon}{4\pi}\bar{M}(x,y) * G(x,y) \tag{C.21}$$

whose Fourier transform is

$$\underline{\bar{F}}(k_x,k_y) = -\frac{\epsilon}{4\pi}\underline{\bar{M}}\,\underline{G}(k_x,k_y) \tag{C.22}$$

The transform of the Green's function is found using a Fourier-Bessel transform [60:12], with the remarkably simple result:

$$\underline{G}(k_x, k_y) = \frac{-j2\pi}{\sqrt{k^2 - k_x^2 - k_y^2}} \qquad \text{(C.23)}$$

Using this with the derivative property (C.16), the MFIE in the spectral domain is

$$\hat{n} \times \underline{\bar{I}} = \frac{1}{j\omega\mu\epsilon}\left\{\hat{x}\left[(k^2 - k_x^2)\underline{E}_x - k_x k_y \underline{E}_y\right] + \hat{y}\left[(k^2 - k_y^2)\underline{E}_y - k_x k_y \underline{E}_x\right]\right\} \qquad \text{(C.24)}$$

Taking the cross product of $\hat{z}$ with (C.24), substituting (C.22) for $\bar{E}$, and using (C.1):

$$\underline{\bar{I}} = \frac{1}{2k\kappa\eta}\begin{bmatrix} (k^2 - k_y^2) & k_x k_y \\ k_x k_y & (k^2 - k_x^2) \end{bmatrix} \cdot \underline{\bar{E}} \qquad \text{(C.25)}$$

$$\kappa = \sqrt{k^2 - k_x^2 - k_y^2} \qquad \text{(C.26)}$$

or in terms of the unit cell electric field:

$$\underline{\bar{I}} = \sum_m \sum_n \bar{\bar{T}} \cdot \underline{\bar{E}}_{uc} \delta(k_x - k_{xm}, k_y - k_{yn}) \qquad \text{(C.27)}$$

$$\bar{\bar{T}} = \frac{1}{2k\kappa\eta}\begin{bmatrix} (k^2 - k_y^2) & k_x k_y \\ k_x k_y & (k^2 - k_x^2) \end{bmatrix} \qquad \text{(C.28)}$$

Finally, testing is to be carried out in the spatial domain, requiring the inverse transform of (C.25). With the delta function in the integrand, that integration reduces to a sampling of the integrand at each k_{xm} and k_{yn}:

$$\bar{J} = \sum_m \sum_n \bar{\bar{T}}_{mn} \cdot \underline{\bar{E}}_{uc}(k_{xm}, k_{yn}) e^{-jk_{xm}x} e^{-jk_{yn}y} \qquad \text{(C.29)}$$

This form of the MFIE is only valid for rectangular array lattices, but many actual phased arrays

$$\ddot{\bar{T}}_{mn} = \frac{1}{2d_x d_y k \kappa_{mn} \eta} \begin{bmatrix} (k^2 - k_{yn}^2) & k_{xm} k_{yn} \\ k_{xm} k_{yn} & (k^2 - k_{xm}^2) \end{bmatrix} \tag{C.30}$$

$$\kappa_{mn} = \left[k^2 - k_{xm}^2 - k_{yn}^2 \right]^{1/2}$$

use triangular lattices. The following section will show how (C.29) is modified to accomodate skewed lattices.

C.3. Skewed Array Lattices

Phased array antenna elements are usually arranged in a triangular lattice, formed by shifting successive rows to the right or left by one half the column spacing. This allows a larger inter-element spacing (hence fewer elements for a given aperture area) to cover a given scan region without grating lobes.

Figure 12 shows an even more general case in which the shift between successive rows is not necessarily one half the column spacing d_x. The Floquet condition for this situation is

$$\bar{E}(x + md_x + nd_y \cot\gamma, y + nd_y) = \bar{E}(x,y) e^{-j\psi_x(md_x + nd_y \cot\gamma)} e^{-j\psi_y nd_y} \tag{C.31}$$

or in convolution form

$$\bar{E}(x,y) = \bar{E}_{uc}(x,y) * \sum_m \sum_n \delta(x - md_x - nd_y \cot\gamma, y - nd_y) e^{-j\psi_x x} e^{-j\psi_y y} \tag{C.32}$$

The Fourier transform of (C.30) is required, but it may be obtained without directly performing the integration. As in the case of the rectangular lattice, a result similar to (C.16) is expected, i.e. the unit cell transform times a series of spectral domain delta functions. The Fourier transform pair (C.9) and (C.16) have a unique interpretation in terms of *direct* and *reciprocal lattices*

116

[61:94-98]. The coordinates $(k_x, k_y) = (2\pi m/d_x, 2\pi n/d_y)$ are points in the reciprocal lattice corresponding to $(x,y) = (md_x, nd_y)$ and $(4\pi^2/d_x d_y)$ is the area of one of its unit cells.

In the skewed lattice, the coordinates of any element are integer multiples of the basis vectors $\bar{a}$ and $\bar{b}$:

$$\begin{aligned} \bar{a} &= \hat{x} d_x \\ \bar{b} &= \hat{x} d_x \cot\gamma + \hat{y} d_y \end{aligned} \tag{C.33}$$

The basis vectors in the reciprocal lattice, $\bar{\alpha}$ and $\bar{\beta}$ are found by solving

$$\begin{aligned} \bar{a} \cdot \bar{\alpha} &= \bar{b} \cdot \bar{\beta} = 1 \\ \bar{a} \cdot \bar{\beta} &= \bar{b} \cdot \bar{\alpha} = 0 \end{aligned} \tag{C.34}$$

with the result

$$\bar{\alpha} = \hat{x} \left[\frac{1}{d_x} \right] - \hat{y} \left[\frac{\cot\gamma}{d_x} \right] \tag{C.35}$$

$$\bar{\beta} = \hat{y} \left[\frac{1}{d_y} \right] \tag{C.36}$$

The spatial frequency coordinates corresponding to points in the reciprocal lattice are

$$k'_{xmn} = 2\pi\hat{x} \cdot (m\alpha + n\beta) = \frac{2\pi m}{d_x} \tag{C.37}$$

$$k'_{ymn} = 2\pi\hat{y} \cdot (m\bar{\alpha} + n\bar{\beta}) = \frac{2\pi n}{d_y} - \frac{2\pi m \cot\gamma}{d_x} \tag{C.38}$$

The primes signify that these are the unscanned lattice points. The unit cell area in a skewed lattice is the same as in a non-skewed lattice with the same d_x and d_y, so the spectral domain unit cell area is also the same. The end result is that the Fourier transform of (C.32) is

117

$$\bar{\bar{E}}(k_x, k_y) = \bar{\bar{E}}_{uc} \cdot \frac{4\pi^2}{d_x d_y} \sum_m \sum_n \delta(k_x - k_{xmn}, k_y - k_{ymn}) \qquad \text{(C.39)}$$

$$k_{xmn} = \frac{2\pi m}{d_x} - k_0 \sin\theta_0 \cos\phi_0 \qquad \text{(C.40)}$$

$$k_{ymn} = \frac{2\pi n}{d_y} - \frac{2\pi m \cot\gamma}{d_x} - k_0 \sin\theta_0 \sin\phi_0 \qquad \text{(C.41)}$$

The effect of beam steering is to shift all of the lattice points. The last three equations above explain the origin of the Floquet harmonics for a skewed lattice, and agrees with the result given by Mittra et. al. [62:1596]. The relationship between $\hat{z} \times \bar{H}$ and $\bar{E}$ is still as given by (C.29), except that the sample points are now k_{xmn}, k_{ymn} instead of k_{xm}, k_{yn}.

C.4. Expansion Function Fourier Transform

The electric field within the unit cell is expanded in known vector functions $\bar{\psi}_s$ with unknown complex scalar coefficients e_s. By linearity of Fourier transforms, the unit cell field is

$$\bar{\bar{E}}_{uc} = \sum_{s=1}^{N} e_s \bar{\xi}_s(k_x, k_y) \qquad \text{(C.42)}$$

where $\bar{\xi}_s$ is the two dimensional Fourier transform of $\bar{\psi}_s$. These may be evaluated analytically with the help of homogeneous coordinates within the triangles subdividing the radiation boundary (the faces of those tetrahedra bordering on the boundary).

Suppose that within triangle k edge s goes between local nodes i and j. Then in terms of the 2D homogeneous coordinates,

$$\bar{\psi}_{ij}^{(k)} = L_{ij}(t_i \nabla t_j - t_j \nabla t_i) \qquad \text{(C.43)}$$

118

Let τ_i denote the Fourier transform of the scalar linear finite element defined at node i, denoted f_i. In the homogeneous coordinates, $f_i = t_i$ and

$$\tau_i = \int\int_{-\infty}^{\infty} t_i \, e^{jk_x x} e^{jk_y y} \, dx \, dy = 2A \int_0^1 dt_2 \int_0^{1-t_2} t_i \, e^{jk_x x} e^{jk_y y} \, dt_1 \qquad (C.44)$$

The factor 2A is the inverse of the Jacobian of the transform from (x,y) to $(t_1, t_2$. The inverse coordinate transform is expressed in terms of six constants given in Appendix B, (B.7),(B.8). Let B_l denote the following combinations:

$$B_l = \alpha_l k_x + \beta_l k_y \,, \quad l = 1,2,3 \qquad (C.45)$$

Then substituting into (C.44):

$$\tau_i = 2A \, e^{jB_0} \int_0^1 e^{jB_2 t_2} \, dt_2 \int_0^{1-t_2} t_i \, e^{jB_1 t_1} \, dt_1 \qquad (C.46)$$

Five separate cases must be considered, depending on the range of the constants B_l.

(a) general case

$$\tau_l = 2A e^{jB_0} \left\{ \frac{-j}{B_1^2 B_2} + \frac{j e^{jB_2}}{B_2 (B_1 - B_2)^2} + \frac{e^{jB_1}(jB_2 - 2jB_1 + B_1 B_2 - B_1^2)}{B_1^2 (B_1 - B_2)^2} \right\} \qquad (C.47)$$

(b) $|B_1 - B_2| << 1$

$$\tau_l = \frac{2A e^{jB_0}}{B_2^2} \left\{ e^{jB_2} \left[\frac{1}{2} - \frac{jB_2}{2} + \frac{jB_1}{3} + \frac{B_2(B_1 - B_2)}{12} \right] + \frac{1}{B_1^2} \left[1 - e^{jB_1}(1 - jB_1) \right] \right\} \qquad (C.48)$$

119

(c) $|B_2| << 1$

$$\tau_1 = \frac{2Ae^{jB_0}}{B_1^4}\left\{ -e^{jB_1}\left[j(2B_1 + 3B_2) + B_1(B_1 + B_2) \right] \right.$$

$$\left. + \left[j(2B_1 + 3B_2 - \tfrac{1}{2}B_1^2 B_2) - B_1(B_1 + 2B_2) \right] \right\} \tag{C.49}$$

(d) $|B_1| << 1$

$$\tau_1 = \frac{2Ae^{jB_0}}{B_2^4}\left\{ E^{jB_2}(jB_2 + 2jB_1) + \right.$$

$$\left. \left[(2B_1 B_2 + B_2^2 - \tfrac{1}{3}B_1 B_2^3) + j(-2B_1 - B_2 + B_1 B_2^2 + \tfrac{1}{2}B_2^3) \right] \right\} \tag{C.50}$$

(e) $|B_1| << 1$ and $|B_2| << 1$

$$\tau_1 = \frac{2Ae^{jB_0}}{120}(20 + 10jB_1 + 5jB_2 - 2B_1 B_2) \tag{C.51}$$

The small argument forms ensure numerical stability. The expressions for τ_2 are obtained by reversing B_1 and B_2.

C.5. Integral Equation from Floquet Modes

The following is a more conventional derivation of the periodic integral equation using "Floquet modes." It follows the same general procedure used in deriving the integral equations for waveguides.

The orthonormal vector Floquet modes [29:41-42] are analogous to waveguide mode functions:

$$\bar{g}_{1mn} = \frac{1}{\sqrt{d_x d_y}}\left[\frac{\hat{x}k_{yn} - \hat{y}k_{xm}}{\sqrt{k_{xm}^2 + k_{yn}^2}} \right] e^{j(k_{xm}x + k_{yn}y)} \qquad \text{(TE)} \tag{C.52}$$

120

$$\bar{g}_{2mn} = \frac{1}{\sqrt{d_x d_y}} \left[\frac{\hat{x} k_{xm} + \hat{y} k_{yn}}{\sqrt{k_{xm}^2 + k_{yn}^2}} \right] e^{j(k_{xm}x + k_{yn}y)} \qquad (\text{TM}) \qquad (\text{C.53})$$

The corresponding modal admittances are

$$Y_{pmn} = \begin{cases} \kappa_{mn}/k_0 \eta_0 & p=1 \quad (\text{TE}) \\ k_0/\kappa_{mn} \eta_0 & p=2 \quad (\text{TM}) \end{cases} \qquad (\text{C.54})$$

Using the waveguide integral equation from Chapter IV (29) but expanding the mode sum over three indices:

$$\sum_{m=0}^{\infty} \sum_{n=1}^{\infty} \sum_{p=1}^{2} Y_{pmn} \bar{g}_{pmn} \int_{\Gamma_R} \bar{E}_t \cdot \bar{g}_{pmn} \, ds - \bar{J} = 0 \qquad (\text{C.55})$$

$\bar{E}_t$ is the transverse (to z) unit cell electric field on Γ_R. Due to the form of the complex exponential factors in (C.52) and (C.53), the integral (C.55) results in factors involving its Fourier transform, $\underline{\bar{E}}_t$. The TE and TM mode terms from (C.55) are

$$Y_{1mn} \bar{g}_{1mn} \int_{\Gamma_R} \bar{E}_t \cdot \bar{g}_{1mn} \, ds = \frac{\kappa_{mn}/k_0 \eta_0}{d_x d_y (k_{xm}^2 + k_{yn}^2)} \cdot$$
$$\left[\hat{x} \underline{E}_x k_{yn}^2 - \hat{x} \underline{E}_y k_{xm} k_{yn} - \hat{y} \underline{E}_x k_{xm} k_{yn} + \hat{y} \underline{E}_y k_{xm}^2 \right] \qquad (\text{C.56})$$

$$Y_{2mn} \bar{g}_{2mn} \int_{\Gamma_R} \bar{E}_t \cdot \bar{g}_{2mn} \, ds = \frac{k_0/\kappa_{mn} \eta_0}{d_x d_y (k_{xm}^2 + k_{yn}^2)} \cdot$$
$$\left[\hat{x} \underline{E}_x k_{xm}^2 + \hat{x} \underline{E}_y k_{xm} k_{yn} + \hat{y} \underline{E}_x k_{xm} k_{yn} + \hat{y} \underline{E}_y k_{yn}^2 \right]$$

Summing the TE and TM modes:

121

$$\sum_{p=1}^{2} Y_{pmn} \, \bar{g}_{pmn} \int_{\Gamma_R} \bar{E}_t \cdot \bar{g}_{pmn} \, ds = \frac{1}{\kappa_{mn} \eta_0 k_0 d_x d_y} \cdot$$

$$\left[\hat{x} \underline{E}_x (k_0^2 - k_{yn}^2) + \hat{x} \underline{E}_y k_{xm} k_{yn} + \hat{y} \underline{E}_x k_{xm} k_{yn} + \hat{y} \underline{E}_y (k_0 - k_{xm}^2) \right]$$

(C.58)

When this is written in dyadic notation, it is clear that the integral equations (B.29) and (B.55) are the same.

Appendix D

Periodic Boundary Conditions
for the Finite Element Problem

This appendix develops the method for applying periodic boundary conditions to a finite element problem in one dimension. Consider the two functions f(x) and h(x) shown in Figure

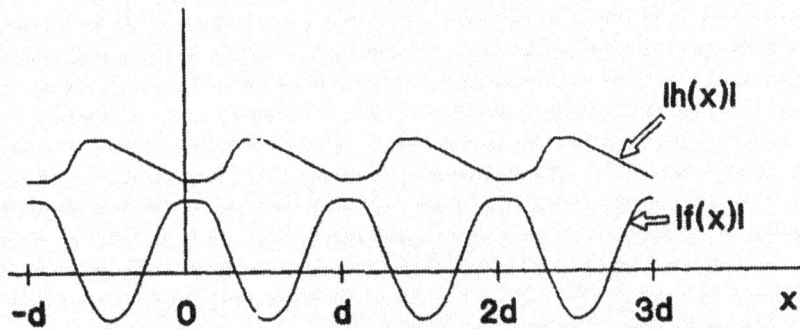

Figure D1. Periodic Functions

D1, related by a linear operator equation Lf=h. Their magnitudes are periodic, repeating on each interval, but each interval has a progressive phase shift relative to the next:

$$f(x+nd) = f(x)e^{j\psi n} \tag{D.1}$$

This may be regarded as a *periodic boundary condition* for the function on the interval [0,d]. For example purposes the method of weighted residuals will be used to produce a functional:

$$F(f) = \langle Lf, w \rangle - \langle h, w \rangle$$
$$= \int_0^d \left[L(f)w^* - hw^* \right] dx \tag{D.2}$$

The functions f, h and w will be represented as sums of complex coefficients (f_i, h_i, and w_i) times scalar basis functions $t_i(x)$ where i may range from $-\infty$ to $+\infty$. N+1 of these expansion

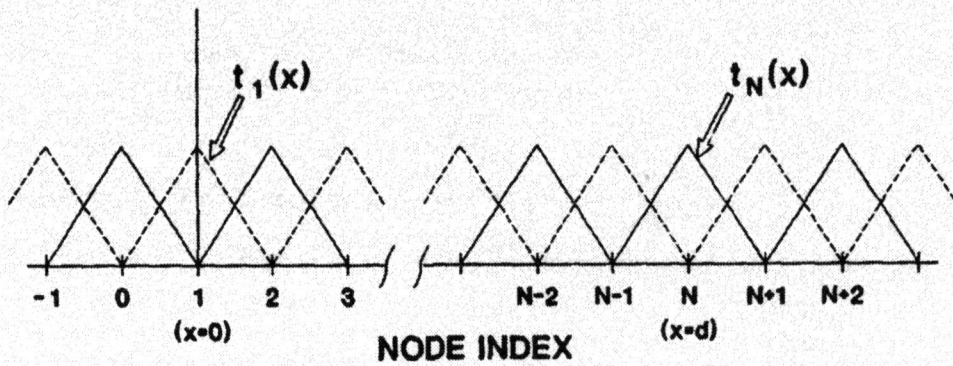

Figure D2. Expansion/Weighting Functions

functions are nonzero within the interval [0,d] as illustrated in Figure D2. f_1 and h_1 are the values $f(0)$ and $h(0)$; and f_N and h_N are the values $f(d)$ and $h(d)$. The functions $t_i(x)$ are not necessarily linear as shown, and do not necessarily extend over subintervals of the same length. However, those in successive intervals must be replicas of each other with $t_{i+N}(x) = t_{i+1}(x-d)$. After expanding the functions in (D.2), the derivative of F with respect to each w_j gives an infinite tridiagonal system with matrix and right hand side entries

$$s_{j,i} = \int_{-\infty}^{\infty} L[t_i(x)] t_j(x)\, dx \qquad (D.3)$$

$$g_j = \int_{-\infty}^{\infty} h(x) t_j(x)\, dx \qquad (D.4)$$

For example, the equations pertaining to the nodes -1,0,1,2,...,N-! 3 are

124

$$\vdots$$

$$
\begin{array}{ccccccc}
s_{0,-1}f_{-1} & + & s_{0,0}f_0 & + & s_{0,1}f_1 & = & g_0 \\
s_{1,0}f_0 & + & s_{1,1}f_1 & + & s_{1,2}f_2 & = & g_1 \\
s_{2,1}f_1 & + & s_{2,2}f_2 & + & s_{2,3}f_3 & = & g_2
\end{array}
$$

$$\vdots \qquad\qquad\qquad\qquad\qquad\qquad (D.5)$$

$$
\begin{array}{ccccccc}
s_{N-1,N-2}f_{N-2} & + & s_{N-1,N-1}f_{N-1} & + & s_{N-1,N}f_N & = & g_{N-1} \\
s_{N,N-1}f_{N-1} & + & s_{N,N}f_N & + & s_{N,N+1}f_{N+1} & = & g_N \\
s_{N+1,N}f_N & + & s_{N+1,N+1}f_{N+1} & + & s_{N+1,N+2}f_{N+2} & = & g_{N+1}
\end{array}
$$

$$\vdots$$

The periodicity conditions on the discrete coefficients f and g are

$$
\begin{array}{ll}
f_{i+N-1} = f_i e^{j\psi} \;\; ; & f_{i-N+1} = f_i e^{-j\psi} \\
g_{i+N-1} = g_i e^{j\psi} \;\; ; & g_{i-N+1} = g_i e^{-j\psi}
\end{array}
\qquad (D.6)
$$

The matrix elements must also satisfy a periodicity condition. Since t_{i+1} and t_{i+N} are identical for all i, it is clear that $s_{j+1,\,i+1} = s_{i+N,\,j+N}$. We can rewrite the system so that it only involves the unknown values of f and known values of g within the interval [0,d):

$$\vdots$$

$$
\begin{array}{ccccccccc}
s_{N-1,N-2}f_{N-2}e^{-j\psi} & + & s_{N-1,N-1}f_{N-1}e^{-j\psi} & + & s_{N-1,N}f_1 & = & g_{N-1}e^{-j\psi} & (a) \\
s_{N,N-1}f_{N-1}e^{-j\psi} & + & s_{1,1}f_1 & + & s_{1,2}f_2 & = & g_1 & (b) \\
s_{2,1}f_1 & + & s_{2,2}f_2 & + & s_{2,3}f_3 & = & g_2 & (c)
\end{array}
$$

$$\vdots \qquad\qquad\qquad\qquad\qquad\qquad (D.7)$$

$$
\begin{array}{ccccccccc}
s_{N-1,N-2}f_{N-2} & + & s_{N-1,N-1}f_{N-1} & + & s_{N-1,N}f_1 e^{j\psi} & = & g_{N-1} & (d) \\
s_{N,N-1}f_{N-1} & + & s_{1,1}f_1 e^{j\psi} & + & s_{1,2}f_2 e^{j\psi} & = & g_1 e^{j\psi} & (e) \\
s_{2,1}f_1 e^{j\psi} & + & s_{2,2}f_2 e^{j\psi} & + & s_{2,3}f_3 e^{j\psi} & = & g_2 e^{j\psi} & (f)
\end{array}
$$

$$\vdots$$

Multiplying (b) and (c) by $e^{j\psi}$ and subtracting from (e) and (f), repsectively, eliminates the latter three from the system. Similarly, multiplying (d) by $e^{-j\psi}$ and subtracting from (a) eliminates (a).

Continuing the process will eliminate all equations preceding (b) and succeeding (d) leaving only a reduced system of N equations:

$$s_{N,N-1} f_{N-1} e^{-j\psi} \quad + \quad s_{1,1} f_1 \quad + \quad s_{1,2} f_2 \quad = g_1$$
$$s_{2,1} f_1 \quad + \quad s_{2,2} f_2 \quad + \quad s_{2,3} f_3 \quad = g_2 \qquad \text{(D.8)}$$
$$\vdots$$
$$s_{N-1,N-2} f_{N-2} \quad + \quad s_{N-1,N-1} f_{N-1} \quad + \quad s_{N-1,N} f_1 e^{j\psi} \quad = g_{N-1}$$

Suppose that now the original problem geometry is truncated at the boundaries $x=0$ and $x=d$, and the two ends are "wrapped" back on each other, as shown in Figure D3. Now nodes

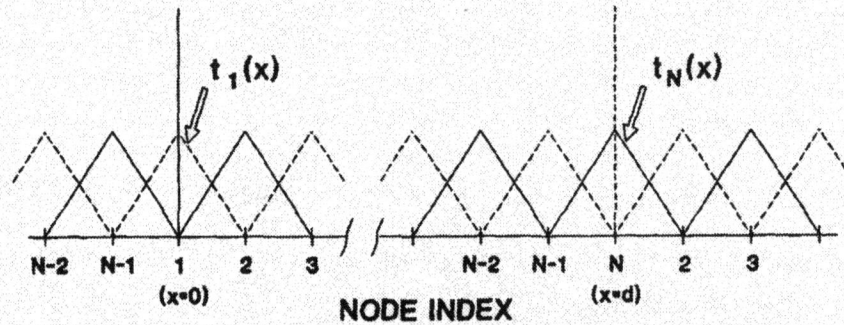

Figure D3. "Wrapped" Domain

1 and N are the same point, as are 2 and N+1, etc. Th n the inner product in (D.3) has new terms for $(i,j)=(1,N-1),(N-1,1)$, and referring to (D.8) it is evident that

$$s_{1,N-1} = s_{N,N-1} e^{-j\psi}$$
$$s_{N-1,1} = s_{N-1,N} e^{j\psi} \qquad \text{(D.9)}$$

This system is no longer tridiagonal because the boundary terms have introduced new elements:

126

$$[S] = \begin{bmatrix} s_{1,1} & s_{1,2} & 0 & 0 & \cdots & 0 & 0 & s_{N,N-1}\,e^{-j\psi} \\ s_{2,1} & s_{2,2} & s_{2,3} & 0 & \cdots & 0 & 0 & 0 \\ 0 & s_{3,2} & s_{3,3} & s_{3,4} & \cdots & 0 & 0 & 0 \\ & & & \cdot & & & & \\ & & & \cdot & & & & \\ & & & \cdot & & & & \\ 0 & 0 & 0 & 0 & \cdots & s_{N-2,N-2} & s_{N-2,N-1} & 0 \\ s_{N-1,N}\,e^{j\psi} & 0 & 0 & 0 & \cdots & s_{N-1,N-2} & s_{N-1,N-1} & s_{N-1,N} \end{bmatrix} \qquad \text{(D.10)}$$

If the original (infinite) matrix was Hermitian (adjoint), then $s_{N,N-1} = s^{*}_{N-1,N}$, and the new system is adjoint as well. This indicates that periodic boundary conditions do not necessarily cause an operator to become non-self-adjoint.

127

References

1. Stark, L., "Microwave Theory of Phased-Array Antennas - A Review," *Proceedings of the IEEE*, **62**, pp. 1661-1701, Dec. 1974.

2. Mailloux, R. J., "Phased Array Theory and Technology," *Proceedings of the IEEE*, **70**, pp. 246-291, Mar. 1982.

3. Schell, A.C., "Trends in Phased Array Development," *Phased Arrays 1985 Symposium Proceedings*, Vol. I, Rome Air Development Center: Hanscom AFB, MA, RADC-TR-85-171, pp. 1-6, Sep. 1985.

4. Lewis, L.R., M. Fasset and J. Hunt, "A Broadband Stripline Array Element," *Proc. IEEE Antennas & Propagation Int'l Symp.*, Atlanta GA, pp. 335-337, Jun. 1974.

5. Cooley, M.E., D.H. Schaubert, N.E. Buris and E.A. Urbanik, "Radiation and Scattering Analysis of Infinite Arrays of Endfire Slot Antennas with a Ground Plane," *IEEE Trans. Antennas Propagat.*, **AP-39**, pp. 1615-1625, Nov. 1991.

6. Edward, B. and D. Rees, "A Broadband Printed Dipole with Integrated Balun," *Microwave Journal*, pp. 339-344, May 1987.

7. Bayard, J-P. R., M.E. Cooley an D.H. Schaubert, "Analysis of Infinite Arrays of Printed Dipoles on Dielectric Sheets Perpendicular to a Ground Plane," *IEEE Trans. Antennas Propagat.*, **AP-39**, pp. 1722-1732, Dec. 1991.

8. Bayard, J-P.R., M.E. Cooley, and D.H. Schaubert, "Effects of E-Plane Electric Walls on Infinite Arrays of Dipoles Printed on Protruding Dielectric Substrates," *IEEE Antennas and Propagation 1992 International Symposium Digest*, pp. 1410-1413, Jul. 1992.

9. Schuman, H.K., D.R. Pflug, and L.D. Thompson, "Infinite Phased Arrays of Arbitrarily Bent Thin Wire Radiators," *IEEE Trans. Antennas Propagat.*, **AP-32**, pp. 364-377, Apr. 1984.

10. Carver, K.R. and J.W. Mink, "Microstrip Antenna Technology," *IEEE Trans. Antennas Propagat.*, **AP-29**, pp. 2-24, Jan. 1981.

11. Herd, J.S., "Full Wave Analysis of Proximity Coupled Rectangular Microstrip Antenna Arrays," *Electromagnetics*, **11**, pp. 21-46, Mar. 1991.

12. Jin, J-M. and V.V. Liepa, "Application of Hybrid Finite Element Method to Electromagnetic Scattering from Coated Cylinders," *IEEE Trans. Antennas Propagat.*, **AP-36**, pp. 50-54, Jan. 1988.

13. Boyse, W.E. and A.A. Seidl, "A Hybrid Finite Element and Moment Method for Electromagnetic Scattering from Inhomogeneous Objects," *Conference on Applied Computational Electromagnetics*, Monterey, CA, pp. 160-169, Mar. 1991.

14. Yuan, Y., D.R. Lynch and J.W. Strohbehn, "Coupling of Finite Element and Moment Methods for Electromagnetic Scattering from Inhomogeneous Objects," *IEEE Trans. Antennas Propagat.*, **AP-38**, pp. 386-393, Mar. 1990.

15. Morgan, M.A., "Principles of Finite Methods in Electromagnetic Scattering," *Progress in Electromagnetics Research: Finite Element and Finite Difference Methods in Electromagnetic Scattering*, pp. 1-68, New York, NY: Elsevier, 1990.

16. Gedney, S.D., J.F. Lee and R. Mittra, "A Combined FEM/MoM Approach to Analyze the Plane Wave Diffraction by Arbitrary Gratings," *IEEE Trans. Microwave Theory Tech.*, **MTT-40**, pp. 363-370, Feb. 1992.

17. McGrath, D.T., "Hybrid Finite Element/Waveguide Mode Analysis of Passive RF Devices," RL-TR-93-, Hanscom AFB ,MA: USAF Rome Laboratory, Feb. 1993.

18. Strang, G. and G.J. Fix, *An Analysis of the Finite Element Method*, Englewood Cliffs, NJ: Prentice-Hall, 1973.

19. Mur, G., "Finite-Element Modeling of Three-Dimensional Electromagnetic Fields in Inhomogeneous Media," *Radio Science*, **26**, pp. 275-280, 1991.

20. Paulsen, K.D. and D.R. Lynch, "Elimination of Vector Parasites in Finite Element Maxwell Solutions," *IEEE Trans. Microwave Theory Tech.*, **MTT-39**, pp. 395-404, Mar. 1991.

21. Boyse, W.E., D.R. Lynch, K.D. Paulsen and G.N. Minerbo, "Nodal-Based Finite Element Modeling of Maxwell's Equations," *IEEE Trans. Antennas Propagat.*, **AP-40**, pp. 642-651, Jun. 1992.

22. Nedelec, J.C., "Mixed Finite Elements in R^3," *Numerische Mathematik*, **35**, pp. 315-341, 1980.

23. Barton, M.L. and Z.J. Cendes, "New Vector Finite Elements for Three-Dimensional Magnetic Field Computation," *J. Appl. Phys.* **61**, pp. 3919-3921, Apr. 1987.

24. D'Angelo, J.D. and I.D. Mayergoyz, "Finite Element Methods for the Solution of RF Radiation and Scattering Problems," *Electromagnetics*, **10**, pp. 177-199, 1990.

25. K.D. Paulsen, W.E. Boyse and D.R. Lynch, "Continuous Potential Maxwell Solutions on Nodal-Based Finite Elements," *IEEE Trans. Antennas Propagat.*, **AP-40**, pp. 1192-1200, Oct. 1992.

26. Silvester, P.P. and R.L. Ferrari, *Finite Elements for Electrical Engineers*, 2nd. ed., Cambridge Univ. Press, 1990.

27. Harrington, R.F. and J.R. Mautz, "A Generalized Network Formulation for Aperture Problems," *IEEE Trans. Antennas Propagat.*, **AP-24**, pp. 870-873, Nov. 1976.

28. Marcuvitz, N., *Waveguide Handbook*, New York: McGraw-Hill, 1951.

29. Amitay, N., V. Galindo and C. Wu, *Theory and Analysis of Phased Array Antennas*, New York: Wiley, 1972.

30. Gaskill, J.D., *Linear Systems, Fourier Transforms, and Optics*, New York: John Wiley & Sons, 1978.

31. IMSL, Inc., *Users' Manual: IMSL MATH/Library - FORTRAN Subroutines for Mathematical Applications*, Dec. 1989.

32. Anderson, E., et. al., *LAPACK Users' Guide*, Philadelphia, PA: SIAM, 1992.

33. Sarkar, T.K. and E. Arvas, "On a Class of Finite Step Iterative Methods (Conjugate Directions) for the Solution of an Operator Equation Arising in Electromagnetics," *IEEE Trans. Antennas Propagat.*, **AP-33**, pp. 1058-1066, Oct. 1985.

34. Potter, P.D. and A.C. Ludwig, "Beamshaping by Use of Higher Order Modes in Conical Horns," *Electromagnetic Horn Antennas*, ed. A.W. Love, pp. 203-204, New York: IEEE Press, 1976.

35. Masterman, P.H. and P.J.B. Clarricoats, "Computer Field-Matching Solution of Waveguide Transverse Discontinuities," *Proc. IEE*, **188**, pp. 51-63, Jan. 1971.

36. Kowalski, G. and R. Pregla, "Dispersion Characteristics of Shielded Microstrips with Finite Thickness," *Arch. Elek. Ubertragung*, **25**, pp. 193-196, Apr. 1971.

37. Wheeler, H.A., "Transmission Line Properties of a Strip on a Dielectric Sheet on a Plane," *IEEE Trans. Microwave Theory Tech.*, **MTT-25**, pp. 631-647, Aug. 1977.

38. Rowe, D.A. and B.Y. Lao, "Numerical Analysis of Shielded Coplanar Waveguide," *IEEE Trans. Microwave Theory Tech.*, **MTT-31**, pp. 911-915, Nov. 1983.

39. Webb, J.P., G.L. Maile, and R.L. Ferrari, "Finite Element Solution of Three Dimensional Electromagnetic Problems," *Proc. IEE*, **130**, pp. 153-159, Mar. 1983.

40. Diamond, B.L., "Resonance Phenomena in Waveguide Arrays," *Proc. 1967 IEEE Antennas Propagat. Int'l Symp.*, Ann Arbor, MI, pp. 110-115, Oct. 1967.

41. Lee, S-W and W. Jones, "On the Suppression of Radiation Nulls and Broadband Impedance Matching of Rectangular Waveguide Phased Arrays," *IEEE Trans. Antennas Propagat.*, **AP-19**, pp. 41-51, Jan. 1971.

42. Amitay, N. and M.J. Gans, "Design of Rectangular Horn Arrays with Oversized Aperture Elements," *IEEE Trans. Antennas Propagat.*, **AP-29**, pp. 871-884, Nov. 1981.

43. Tang, R. and N.S. Wong, "Multimode Phased Array Element for Wide Scan Angle Impedance Matching," *Proc. IEEE*, **56**, pp. 1951-1959, Nov. 1968.

44. Structural Dynamics Research Corporation, "Integrated Design Engineering Analysis Software Users' Manual," S.D.R.C., Milford, OH, 1990.

45. Pozar, D.M. and D.H. Schaubert, "Analysis of an Infinite Array of Rectangular Microstrip Patches with Idealized Probe Feeds," *IEEE Trans. Antennas Propagat.*, **AP-32**, pp. 1101-1107, Oct. 1984.

46. Fenn, A.J., "Theoretical and Experimental Study of Monopole Phased Array Antennas," *IEEE Trans. Antennas Propagat.*, **AP-33**, pp. 1131-1142, Oct. 1985.

47. Herper, J.C. and A. Hessel, "Performance of $\lambda/4$ Monopole in a Phased Array," *Proc. 1975 Antennas and Propagation Int'l Symp.*, Urbana, IL, Jun. 1975.

48. Hammerstad, E.O., "Equations for Microstrip Circuit Design," *Proc. 5th European Microwaves Conf.*, Hamburg, pp. 268-272, 1975.

49. Gupta, K.C., R. Gharg, and R. Chadha, *Computer-Aided Design of Microwave Circuits*, Dedham, MA: Artech House, 1981.

50. Schaubert, D.H., "Endfire Slotline Antennas," *Proc. JINA '90*, 253-265, Nov. 1990.

51. Choung, Y.H. and C.C. Chen, "44 GHz Slotline Phased Array Antenna," *Proc. 1989 Antennas and Propagation Int'l Symp.*, pp. 1730-1733.

52. Ho, T.Q. and S.M. Hart, "A Broad-Band Coplanar Waveguide to Slotline Transition," *IEEE Microwave and Guided Wave Letters*, **2**, pp. 415-416, Oct. 1992.

53. Jin, J-M. and J.L. Volakis, "A Hybrid Finite Element Method for Scattering and Radiation by Microstrip Patch Antennas and Arrays Residing in a Cavity," *IEEE Trans. Antennas Propagat*, **AP-39**, pp. 1598-1604, Nov. 1991.

54. Mikhlin, S.G., *Variational Methods in Mathematical Physics*, New York: MacMillan, 1974.

55. Harrington, R.F., *Time-Harmonic Electromagnetic Fields*, New York: McGraw-Hill, 1961.

56. Naylor, A.W. and G.R. Sell, *Linear Operator Theory in Engineering and Science*, New York: Springer-Verlag, 1982.

57. Stroud, A.H. and D. Secrest, *Gaussian Quadrature Formulas*, New York: Prentice-Hall, 1966.

58. Abramowitz, M. and I.E. Stegun, *Handbook of Mathematical Functions*, National Bureau of Standards, 1972.

59. Collin, R.E., *Field Theory of Guided Waves*, 2nd Ed., IEEE Press, 1991.

60. Goodman, J.W., *Introduction to Fourier Optics*, McGraw-Hill, 1968.

61. Brillouin, L., *Wave Propagation in Periodic Structures*, New York: Dover, 1953.

62. Mittra, R., C.H. Chan and T. Cwik, "Techniques for Analyzing Frequency Selective Surfaces--A Review, *IEEE Proceedings*, **76**, pp. 1593-1615, Dec, 1988.